How to Get More Miles Per Gallon
Per Gallon
in the 1990s

Robert Sikorsky

TAB BOOKS
Blue Ridge Summit, PA

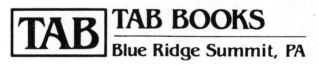

FIRST EDITION
FIRST PRINTING

© 1991 by **TAB Books**.
TAB Books is a division of McGraw-Hill, Inc.

Library of Congress Cataloging-in-Publication Data

Sikorsky, Robert.
    How to get more miles per gallon in the 1990s / by Robert Sikorsky.
       p.   cm.
    Includes index.
    ISBN 0-8306-3793-1 (p)   ISBN 0-8306-8793-9 (h)
    1. Automobiles—Fuel consumption.   I. Title.
TL151.6.S55   1991
629.25′3—dc20   90-29288
                                            CIP

TAB Books offers software for sale. For information and a catalog, please contact TAB Software Department, Blue Ridge Summit, PA 17294-0850.

Questions regarding the content of this book should be addressed to:

Reader Inquiry Branch
TAB Books
Blue Ridge Summit, PA 17294-0850

Acquisitions Editor: Kimberly Tabor
Book Editor: April D. Nolan
Production: Katherine G. Brown
Book Design: Jacyln J. Boone   (TAB1)
Cover Design: Lori E. Schlosser
Cover Illustration: Brian K. Allison

The mileage improvement you get from your car will be directly related to how closely you follow the gas saving tips in this book. By minimizing rolling resistance and aerodynamic drag, increasing engine efficiency, and using economy driving techniques, Shell Oil Company Mileage Marathon drivers coaxed a car almost 400 miles on a single gallon of gas! This book covers all three of these factors, along with many others, in easy-to-read, nontechnical language. Driving and parking techniques; how, when, and where to buy gasoline; practical and workable gas-saving additions and options; fuel conserving alterations, adjustments, and inspections—all play vital roles in improving gas mileage.

But you, the driver, are the single most important factor. Only you can make the conscious decision to drive economically. By using this book as a guide, you'll be well on your way to becoming an expert economy driver. In these days of skyrocketing gasoline prices and the threat of another gas shortage, it is comforting to know you're squeezing every possible mile from each gallon. The effort you make will not only increase gas mileage, conserve fuel, save money, and minimize car repair bills, it will also lessen pollution, save lives, and speed America on its way to being energy self-sufficient.

Introduction
to the 2nd Edition

I wrote the introduction to the first edition of this book in 1978. Isn't it amazing how little things have changed since then? Some of the numbers are different. There are now 145 million cars and 45 million trucks and buses on the road. And we are also using more gasoline (679 gallons per/vehicle), even though our vehicles on average are getting much better fuel economy than they were back then (15.6 mpg average for all motor vehicles).

But conserving fuel and getting more miles per gallon are in vogue—again. Our desire for larger, more powerful, gas-thirsty vehicles has been temporarily put on the back burner—again. Sales of more economical cars is on the rise—again. Blown by rumors of war, we are like reeds in the wind, doing a 180-degree turn toward fuel efficiency and conservation—again.

The most recent gas "crisis," like the past two, has its roots in the Middle East. One of the results has been a jump in the price of gasoline at the pump. It's déjà vu, all over again.

Over the years, I have consistently offered the readers of my syndicated column fuel- and car-saving information, regardless of whether there was a "crisis" or not. Conservation is smart *all* the time and shouldn't be dusted off only in emergencies.

And, in a society that has become more aware of the fragile nature of our environment, it is important to realize that there is a direct relationship between better fuel economy and the amount of carbon dioxide released by a vehicle into the atmosphere. The better the fuel economy, the less globe-warming CO_2 the vehicle produces. You help clean up the air and cool the planet by driving efficiently.

Why many Americans, once a crisis has passed, go right back to inefficient and wasteful ways of driving and maintaining their cars is a mystery to me. Brought up in an atmosphere where nothing was thrown away, I find it hard to comprehend how wasteful some of us are when it comes to our vehicles.

I don't know what will happen to gasoline supplies and prices in the near or distant future. But the fuel-conserving ideas presented in this book, if followed, will greatly improve each vehicle's miles per gallon, lessen our nation's dependence on foreign oil sources, save us money, clean up our air, help us be safer and better drivers, and extend the useful life of our cars. We shouldn't need a crisis to prod us to drive and maintain for economy. It's just plain smart to do so.

How to use
this book

At first glance, well over 300 ways to save gasoline may seem a bit overwhelming, but you will soon see how one suggestion readily relates to another. Each tip pinpoints a specific mileage improvement—whether it is a gas saving addition, an economy driving method, or an easy, do-it-yourself adjustment.

Not all tips apply to all cars. Some are applicable only to older pre-computer vehicles; others are newer-car specific. Most apply to all vehicles.

Read through the book once from beginning to end to get a feel for its scope. Then go back and mark the parts that are of special interest to you. These suggestions will make up your basic program, your starting point in the quest for better mileage. Incorporate them into your driving, and they will soon become habits.

Keep the book handy—the glove compartment of your car is an ideal spot—and refer to it often. Browse through a few pages while parked or when confronted with unavoidable delays. Pick out a couple of techniques and practice them as soon as you get moving, while they're still fresh in your mind.

Commit yourself to fuel conservation and be aware of your driving at all times. Write the words THINK ECONOMY on a piece of paper and tape it to your dashboard. This will act as a constant reminder.

Check your gas mileage with each fill-up. As you see the mpg increase, you will be encouraged to add even more conservation measures to your driving regime.

Chapter **1**

Driving techniques

A race-car driver's ultimate aim is to cross the finish line first, and to do that he or she must drive fast. But a racer must still drive in a way that conserves fuel and the car.

The consistent winners, the ones who collect the big paychecks, are also the ones who seem to drive in an unspectacular fashion. They are intent on conserving their cars and fuel and not making unnecessary maneuvers on the way to the checkered flag. You should follow their lead, even on the way to the grocery store, and drive conservatively, also.

Three-time world-champion race-car driver Jackie Stewart says, "My overruling passion has always been to drive as spectacularly as I can in an unspectacular fashion." True professionals drive so that no one notices them.

"Light is right," says Stewart, referring to a driver's touch on the accelerator and steering wheel. Those three words are important keys to extending the life of your car and getting maximum miles per gallon. It's right if it feels light.

Stewart stresses that the greatest racing drivers are not the ones who manhandle their cars or rip around corners, slipping and sliding, slamming gears up and down and roaring the engine. Almost every successful driver relies on one main driving technique: finesse.

Good drivers use finesse when applying the throttle or brakes. They know how to approach a curve, when to brake before entering it, and when to accelerate out of it. They are gentle with the steering wheel. They respect their cars and know exactly when to upshift or downshift. You should do the same. If the steering wheel begins to feel heavy and cumbersome, you are doing something wrong.

Jackie Stewart says we should strive "to drive like the world's finest chauffeur." Finesse and gentleness should be an integral part of everyday driving. Back-seat passengers should be able to drink a cup of coffee and read the paper without being unduly aware of your stops and starts. You don't have to be a race-car driver to drive this way. This chapter will show you how to drive conservatively, with lightness and finesse, and will put you well on your way to getting every possible mile per gallon from your vehicle.

Although the computers and electronic sensors that control today's newer engines are marvels of high technology, it is still the driver's right foot that plays the major role in getting more miles per gallon. Punch the accelerator, and the car will respond with a burst of speed—regardless if the engine is computer-controlled or not. Those bursts of speed, necessary or not, still cost you gasoline, just like they did in the old days. Conscious control of the attitude of the right foot is what this chapter is all about.

ON THE HIGHWAY

Slow down! Keep your top speed under 55 mph if you want good gas mileage. Excessive speed will put holes in your gas-money pocket. Most cars give best mileage at about 35 to 45 mph. In this speed range, engine efficiency is at its maximum, and wind and rolling resistances are relatively negligible. For every mile per hour over 45, wind resistance increases proportionately, and gas economy suffers. Have you ever had a 60-mph wind blowing in your face? It would knock you down. That is exactly what a moving vehicle must overcome. A car traveling at 60 mph in still air is encountering the wind-resistance effect of a 60 mph headwind on a motionless car!

Depending on car size and frontal area, it takes 10–15 percent more gas to travel at 45 mph than at 35 mph, and another 10–15 percent to speed up from 45 mph to 55 mph. Above 55 mph, mileage suffers even more severe penalties. A car moving at 70 mph gets only two-thirds the gas economy of one going 45 mph—a heavy penalty to pay for excessive speed.

To further dramatize the effects of speed and wind resistance, think about this. A large van with approximately 100 square feet of frontal area requires over 100 extra horsepower to overcome wind resistance at 70 mph, as compared to the same van traveling at 50 mph. Figure 1-1 illustrates speed vs. gas-mileage. The wind resistance, or *aerodynamic drag*, depends upon the speed and shape of the vehicle. The more streamlined a car's body is and the less frontal area it has, the less aerodynamic drag. This aerodynamic drag is represented in a factor called *coefficient of drag* or Cd. The lower the Cd value, the more efficient the car is at higher speeds. Figure 1-2 gives the Cd of different-shaped cars.

Obey the national 55 mph speed limit and you will save gas and contribute to highway safety at the same time. The speed limit was made law so we could preserve our nation's dwin-

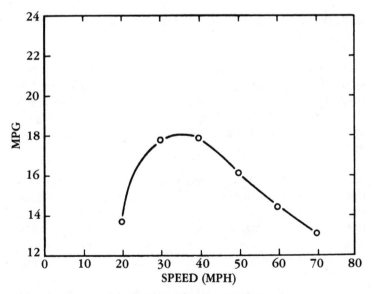

1-1   Road load fuel consumption.

	PERIOD	C_D (APPROX.)
	Late 1920s Early 1930s	0.70
	Late 1930s EArly 1940s	0.58
	Late 1940s Early 1950s	0.52
	Late 1950s Early 1960s	0.50
	1970	0.47
	1980s, Early 1990s	0.33

I-2   Air drag (Cd) of typical cars.

dling petroleum reserves. Obey it, and you'll do your country and yourself a service.

You pay in hard cash and in gasoline burned for every mile over 55 mph you travel. In one study, a test vehicle got 17.3 mpg at 55 mph, and 14.2 mpg at 70 mph. The difference in fuel economy between 70 mph and 55 mph was 3.1 miles per gallon. That represents about $2.60 *per hour* wasted (with gas at $1.50 per gallon).

Drive with the windows closed whenever you can. Open windows, especially at high speeds, create more wind turbulence and have the effect of holding back the car. It takes extra gas to overcome wind drag; in fact, at highway speeds, open windows can lower mileage by as much as 10 percent (FIG. 1-3).

Open windows at highway speeds increase an average car's Cd by about .08. That has the effect of setting back the vehicle's aerodynamics by 30 years. In fact, tests have shown

Contents

Dedication

I originally dedicated this book in 1978 with these words: "For my son Kyle, and all the children of his generation, with the hope that we have the foresight to conserve fuel today so they may know the pleasure and responsibility of driving a car in the future."

Thirteen years have now passed and Kyle and the children of his generation are getting their driver's licenses. It is still my hope that we will learn to be efficient and responsible drivers and conserve both fuel and the planet we live on.

"There is no energy policy that will do as much as voluntary conservation. . . Conservation is our cheapest and cleanest energy source. It helps to control inflation, and every barrel of oil we save is a barrel we don't have to import."

Jimmy Carter

"Nature never gives anything away. Everything is sold at a price."

Ralph Waldo Emerson

Acknowledgments

I wish to thank the following companies and agencies that have all contributed in some way to this book: Shell Oil Company, General Motors Corporation, Ford Motor Company, Saab Cars USA, Inc., McDonnell Douglas Corporation, The American Petroleum Institute, Society of Automotive Engineers, U.S. Department of Energy, Arizona Energy Office, U.S. Department of Transportation, U.S. Environmental Protection Agency, Motor Vehicle Manufacturers Association, Champion Spark Plug Company.

I would also like to thank the New York Times Syndication Sales Corporation for permission to reprint in whole or in part a number of my syndicated columns.

I'm also grateful to Dave Malcheski, former head of the Department of Energy's Driver Awareness Project in Las Vegas, Nevada for his helpful comments and suggestions on the original edition.

Introduction
to the 1st Edition

Unfortunately, there is no panacea for the current high price of gas, but *How to Get More Miles Per Gallon* is a step in the right direction. It's a book for all drivers. By following even a few of the suggestions here, you can dramatically increase your gas mileage. Each paragraph reveals a specific mileage aid. Some will save you gallons of gasoline—others, only a drop or two.

The United States, with only 6 percent of the world's population, uses over 35 percent of the world's energy. This alarming consumption has made us heavily dependent upon foreign oil sources to meet our demands. We currently import approximately 50 percent of the total petroleum we use. Transportation is the single largest use of petroleum in the country, accounting for 50 percent of the total consumed—the equivalent of the entire amount of oil we import!

The average driver gets less than 13.7 mpg from his car and uses somewhere between 600 – 700 gallons of gasoline per year. The 110 million registered automobiles, 30 million registered trucks and 500,000 buses consume a staggering 300 million gallons of gasoline and diesel fuel every day. If the average automobile fuel economy could be improved by only 15 percent, we would cut 28,000,000 gallons per day from that total. Another 8,400,000 gallons daily could be subtracted if every driver who now ignores the 55 mph speed limit were to obey it. If we were to add only one person to each individual commuter's passenger load, an additional savings of 35,000,000 gallons per day would become a reality. These are goals that can be achieved right now.

At 55 MPH
- **WINDOWS OPEN REQUIRE 2 HP FOR DRAG**
- **AIR CONDITIONER REQUIRES ONLY 1 HP**

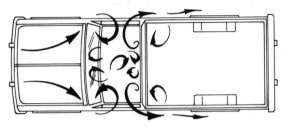

I-3 Air conditioning vs. wind resistance.

that about two horsepower are required to overcome the drag created by open windows at highway speeds. A modern air conditioner, however, uses about one horsepower and, in this case, it is more fuel efficient at highway speeds to use the A/C moderately rather than lowering the windows. However, even though you will save fuel in this instance, you ultimately pay for using the A/C. See Chapter 8 to see how much A/C use can detract from good gas mileage.

Each car has a particular highway speed at which it gets maximum economy. If you have a trip computer that gives instantaneous mpg readings, it's easy to find out what the most efficient speed is for your car. Once you've found it, try to travel as much as possible at or near that number. Each mile you do it means better overall mpg.

When driving on a two-lane highway with a slow-moving vehicle ahead, don't play hide-and-seek behind it. Darting in and out looking for a chance to pass destroys momentum, uses brakes unnecessarily, and plays havoc with mileage. If you see a slow-moving car ahead, ease off the accelerator and decelerate to within a safe distance of it. It's easier and safer to check for oncoming traffic when you keep a respectable distance behind and the view isn't obstructed. Once the way is clear, apply steady pressure to the gas feed and pass. It's cheaper and safer to pass this way; you'll also avoid a great deal of wear and tear on your nerves.

Recent tests involving long-haul truck drivers show a definite relationship between driving fatigue and poorer gas mileage. Make sure you get plenty of rest. Don't drive past your physical capabilities, and you will drive safely and economically.

When on a long trip, drive in your stocking feet occasionally. This gives you a better feel of the accelerator pedal and also keeps you alert.

A very valuable method used by economy-minded drivers to improve highway mileage is to accelerate slowly to a point a few miles under your desired speed. Then ease off the gas until the car reaches that speed, and hold it there with steady pressure. If you have a vacuum gauge you will see the needle rise, indicating better economy, as the car is eased into its cruising speed. In racing language this is known as *maintaining speed with minimum throttle*.

A good race-car driver is as much concerned about conserving fuel as he or she is about staying in front of the pack. That's why *drafting* is a favorite method used to save fuel. By staying close to the car directly in front, a driver can benefit by driving in the vacuum the leading vehicle creates. Although it's not a good idea to get in the habit of drafting (in everyday language, you could call it tailgating!) on public roads, in an emergency it could be helpful.

CITY DRIVING

City driving is composed of 16 percent idling, 35 percent moving at the desired speed, 31 percent accelerating, and 18 percent decelerating. The following section will show you how to squeeze the most miles per gallon out of every city driving mode.

The Environmental Protection Agency (EPA) says that travel habits in the U.S. contribute to poor fuel economy. U.S. autos accumulate about 15 percent of their mileage in trips of 5 miles or less. However, these trips consume more than 30 percent of the nation's automotive fuel because autos operate so ineffici-

ANNUAL FUEL COSTS CHART
For 1991 Model Year
Based on 15,000 Miles per Year

Est MPG	Dollars Per Gallon					
	1.90	1.70	1.50	1.30	1.10	1.05
50	570	510	450	390	330	315
49	582	520	459	398	337	321
48	594	531	469	406	344	328
47	606	543	479	415	351	335
46	620	554	489	424	359	342
45	633	567	500	433	367	350
44	648	580	511	443	375	358
43	663	593	523	453	384	366
42	679	607	536	464	393	375
41	695	622	549	476	402	384
40	713	638	563	488	413	394
39	731	654	577	500	423	404
38	750	671	592	513	434	414
37	770	689	608	527	446	426
36	792	708	625	542	458	438
35	814	729	643	557	471	450
34	838	750	662	574	485	463
33	864	773	682	591	500	477
32	891	797	703	609	516	492
31	919	823	726	629	532	508
30	950	850	750	650	550	525
29	983	879	776	672	569	543
28	1018	911	804	696	589	563
27	1056	944	833	722	611	583
26	1096	981	865	750	635	606
25	1140	1020	900	780	660	630
24	1188	1063	938	813	688	656
23	1239	1109	978	848	717	685
22	1295	1159	1023	886	750	716
21	1357	1214	1071	929	786	750
20	1425	1275	1125	975	825	788
19	1500	1342	1184	1026	868	829
18	1583	1417	1250	1083	917	875
17	1676	1500	1324	1147	971	926
16	1781	1594	1406	1219	1031	984
15	1900	1700	1500	1300	1100	1050
14	2036	1821	1607	1393	1179	1125
13	2192	1962	1731	1500	1269	1212
12	2375	2125	1875	1625	1375	1313
11	2591	2318	2045	1773	1500	1432
10	2850	2550	2250	1950	1650	1575
9	3167	2833	2500	2167	1833	1750

For those who purchase fuel by the liter, 3.785 liters equals one gallon.

I-4 Fuel costs, in dollars, per 15,000 miles.

ently in short trips. Let's turn now to city-driving techniques and see what we can do to cut into that 30 percent figure.

Any car *momentum* (forward movement) you can conserve is better than none. It takes up to six times as much gas to get a car from a dead stop than from a moving speed of just a few miles per hour. Cutting down on the number of times you have to stop is a major requirement for becoming a good economy driver. By eliminating as many complete stops as possible you can get 10 to 25 percent better gas mileage. Preserving momentum is an absolute must and is especially important for city driving. Your gas dollars pay for your car's momentum, so don't waste them. Any maneuver that conserves momentum saves gas.

A 3,500-pound car accelerating from 0 to 35 mph in half a block uses about .025 gallons of gas. The next .025 gallons would carry that same car about half a mile at a steady 35 mph. It costs more to accelerate than it does to keep a car rolling.

Smooth, minimum brake use is a must. A typical driver will use his brakes about 50 percent more than is actually necessary. And each time the brake is applied, precious momentum is lost, fuel is wasted, and brake shoes or pads are worn. Every time you use your brakes, it costs you gasoline to regain the momentum you had before application. Be selective about using your brakes. If you drive ahead and drive defensively, you will find yourself resorting to the brake pedal much less often.

Deceleration is not the same thing as coasting. Coasting is when your foot is off the gas and the car is *out* of gear; deceleration happens when your foot is off the gas while the car is still *in* gear. We decelerate many times each day. But did you know that deceleration is the time when a car gets its highest possible gas mileage?

About 18 percent of all city driving is spent decelerating. Long decelerating approaches to mandatory stops (car in gear, foot off gas) can yield up to 100 mpg for the period the car decelerates. It's a valuable method to use to build up your miles per gallon. Don't rush up to stop signs or traffic lights and then slam on the brakes. Use a long, smooth decelerating approach and watch your mpg start to climb. If you have a trip computer that gives instant miles-per-gallon readings, watch the mpg skyrocket every time you decelerate.

Every state has school zones with posted speed limits requiring a driver to slow down and stop for children. Avoid these when you can and eliminate yet another gas trap.

If a traffic light turns red as you approach it and you know that it will be red when you reach it, ease off the gas and decelerate up to the light as slowly as possible. Even though you will have to stop, a slow, decelerating approach uses the gas carrying you to the light that would have otherwise been wasted idling at the intersection. The slower you approach, the more gas you will save. Use this method any time you have a mandatory stop ahead.

Freeway rush-hour traffic can be likened to an accordion, stretching out and compressing over and over again. Cars creep along at a snail's pace, stop, speed up a bit, stop again, and so on. This continues until traffic thins and normal speeds are resumed at the city's outer limits. Try to keep your speed steady—no matter how slow—and avoid as many stops as possible by keeping some distance between yourself and the car ahead. This extra space acts as a cushion and gives you room to coast. Remember, any momentum, even as little as 2−3 mph, is much better than having to start from a dead stop. Turn off the engine or take the car out of gear during long, unavoidable delays.

According to a recent EPA Emissions and Fuel Economy Report, it takes 50 percent more gasoline to drive under urban conditions than on open highways. This is why mileage figures published for various new cars often show dramatic differences between city and highway fuel economy. To obtain maximum mileage in city conditions, you must know the timing of the traffic lights. There are probably many lights in your own town that you pass countless times each day. Why be trapped by these lights day after day and let them rob you of the mileage you should be getting? With a little effort, you can find out how long the lights stay red or green. Glance at the second hand on your watch or count slowly to yourself as one changes. Most lights operate in the 15-second to one-minute range, although some might be longer. By knowing how long a light stays red or green, you can slow or increase your speed accordingly.

Say you are a block from an intersection and you see the light turn red. From previously having timed the light, you know it stays red for 15 seconds, so you immediately slow down a bit and pace yourself so that the light is green by the time you arrive at the intersection. No lost momentum here. Practice this and soon you will find yourself timing lights everywhere, making fewer stops. You'll be able to use this method countless times, each time deriving satisfaction from the knowledge that you just saved a good deal of gas.

By timing lights you get to where you're going just as fast as the guy who guns his motor at the green only to be stopped a block or two ahead by another red. The drivers in the old Mobilgas Economy Run realized how critical stoplights and traffic patterns were to good mileage and would spend days timing lights and observing traffic along the proposed city section of the route. Then, on the day of the Economy Run, they could zip through the city without stopping and gain those extra miles per gallon needed to win.

Just as avoiding stoplights saves gas, eliminating turns will do likewise. Drive "as the crow flies" for best economy. If two routes are the same in every respect except that one has more turns, the route with the least number of turns will be the most economical. Once the car is moving, its most effortless direction is straight ahead. Plan your trips about town to eliminate as many turns as possible. Each time you turn, valuable momentum is lost that can be regained only by burning more gas.

An economical driver is a skilled driver. If you have problems parking or maneuvering in close quarters, practice to overcome them. It will pay off in a lifetime of gas savings and in peace of mind.

Group as many errands and stops into a single trip as you can. It's better to do more during one trip, while the car is warm, than to make one or two extra trips with a cooler engine. Consolidate your shopping whenever possible.

As you approach an intersection to make a left turn, try to time your approach so that all of the oncoming traffic has passed by the time you begin your turn. In many cases, this will eliminate an unnecessary stop.

Although city driving consumes much more gas than highway driving, it does offer you one way to drive at maximum economy potential, something highway driving can't do. The trick is simply this: Maintain speeds as close as possible to the economical 35–45 mph range. Obviously, on the open highway you can't drive at 35 mph, but in the city there are countless times when you can.

Look at the mileage results of a typical car tested:

Miles Per Hour	Miles Per Gallon
10 mph	4 mpg
20 mph	8 mpg
30 mph	12 mpg
42 mph	16 mpg

At speeds above 42 mph, the fuel economy for this car fell back below 16 mpg, so 42 mph was this particular car's maximum fuel-economy speed.

Keep your car moving in the economy range whenever possible. Although it seems to contradict logic, it does take significantly less fuel to travel at 35–45 mph than it does to travel at 20 mph. Get your car up and over the 20-mph mark when you can. Each mile/per/hour closer to 35 mph means increased gas economy. Creeping does not save gasoline. In fact, slow speeds can accelerate engine-deposit buildup in addition to costing miles per gallon.

Take advantage of the metal crossbars placed in the road near many traffic-light intersections. When activated by the weight of your passing car, the bars (sometimes hidden beneath the road surface) trigger mechanism that causes the light to change from red to green. Know where these are located and you can slow down as you cross them, allow the light to change to green while you are still moving, and continue on through the intersection without stopping. It's another effective way to conserve forward momentum and save gas.

On four-lane (two each way) streets, where an additional lane is provided at intersections for left turns, the best bet for stop-free driving is usually the one closest to the center or passing lane. The left-turn lanes at intersections funnel off left-turn traffic and keep the center lanes moving at a steadier pace than the curb lane, which must slow down or stop for buses, parking cars, taxis, passenger pickup, and right-turn traffic. In city rush-hour madness the center lane will get you home faster and cheaper.

Here's a gas-saving driving technique you'll benefit from a thousand times each year. When approaching a stop sign where cars are already stopped, gauge your speed so that the cars have passed through the sign by the time you arrive. By doing this you will have to stop only once—instead of stopping, pulling up as one car leaves, stopping a second time, pulling up again as the second car leaves, and so on. Slow stop-and-go eats gas, and this is one excellent way to beat it.

When you see a traffic light turn green a block or so ahead and you know (because you now time the length of the lights, as every economy-minded driver should do) it will turn red at about the time you arrive at the intersection, prudent application of the accelerator is in order. You will use less gas by making it through the green light than by stopping at the red and having to start over. Don't gun the car, and stay within specified speed limits—or you could nullify any savings by getting a speeding ticket.

For better economy in stop-and-go traffic on extremely hot days, place the transmission in neutral whenever you are stopped. It eases the strain on the already overworked transmission and prevents it from overheating—the number-one enemy of every transmission. The engine will benefit, too, because it won't have the added burden of turning the idling transmission.

Have correct change ready before stopping at a toll booth or before leaving a pay parking lot. Have letters stamped and ready for mailing at curbside boxes. There are instances galore, similar to these, where a little planning will cut down on engine-idling time. Remember: If the car isn't moving and the engine is running, you're getting 0 mpg.

All states now have a right turn on red law where making right turns on a red light are legal. Take full advantage of this fuel-conserving law. As soon as the way is clear, make your turn. Don't waste gas waiting for the light to turn green.

Don't drive into obvious gas traps. If you want to make a turn at an upcoming intersection but see a long line of cars already waiting, turn a block or two sooner and reroute around an obvious gas-wasting situation. How many times each day do you find yourself stuck at problem intersections? With a little foresight, you can turn sooner, reroute for a few blocks, and avoid these irritating and costly traps. Remember to reroute when circumstances ahead dictate it.

If yours is a two-car family—and almost every family is nowadays—use the car that is already warmed up when running about-town errands. There's no sense in warming up the sec-

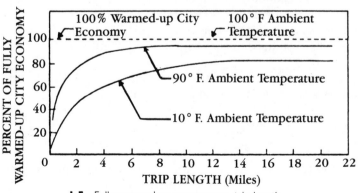

I-5 Fully warmed-up economy vs. trip length.

ond car and paying through the nose for cold-engine gas consumption when you can use the warm, more economical car. This is an especially valuable practice in freezing climates where an engine never becomes fully warmed on short trips. Also, in the already-warmed car, the heater will function faster.

You will enhance the engine life of both cars by using the warm one. You will get better fuel economy with the warm car even though it may have a larger engine than the cold one. In addition, the warm car suffers little engine wear and creates a lot less pollution than if you were to start up the cold-engine vehicle. Figure 1-5 shows that with an outside temperature of 10 degrees Fahrenheit, a car must travel at least 14 miles to attain maximum warmed-up city economy. On short trips of one or two miles, a car may get less than 10 percent of the mileage it can get when it is fully warm.

From the viewpoint of fuel saved, cutting down on short trips of one mile or less is where the big savings come in. The longer the trip, the more efficiently the vehicle will perform. Any time we can use alternate transportation—or walk—for those very short trips, the savings in fuel is compounded.

In a two-car family, always try to use the *most economical vehicle* whenever you can. (One exception is the warm-engine scenario.)

An angry or upset driver isn't a safe driver, nor is he or she a very good economy driver. Stay away from the wheel if you aren't in a good mood. Indeed, this may be the perfect time to try public transportation or use a bicycle. The exercise may be just what you need to get you out of your foul mood.

Drive ahead. Anticipate traffic situations. Be alert. All contribute to better mileage and automotive safety.

ECONOMY TRICKS
FOR ALL DRIVING CONDITIONS

Before getting into your car, look around. Check in front, in back, and to the sides of your vehicle. By doing this you get an

Due to the fuel economy effects of trip length, a typical family car can take the following trips on 25 gallons of gas:

Ten 40-mile trips, or
Sixty 4-mile trips, or
Ninety 2-mile trips, or
One hundred 1-mile trips.

At 1.50 per gallon, the fuel costs would be:

9.4¢ per mile on 40-mile trips, or
15.6¢ per mile on 4-mile trips, or
20.8¢ per mile on 2-mile trips, or
37.6$ per mile on 1-mile trips.

1-6   Cost to the individual motorist.

advance scouting report on how easy or hard it will be to get the car moving, and you can cut down on the time spent looking around after you are in the car and the engine is running.

Before you start your engine, make sure you are ready to go. Adjust seats, light cigarettes, put packages in the back seat, strap the kids in, and do everything you must do before you turn that key.

EPA tests have shown that it is more economical to turn the engine off rather than let it idle if the idle time exceeds 60 seconds. In other words, if the car is going to idle longer than 60 seconds, you save gas by turning the engine off and then restarting it when ready. If the engine is going to idle less than 60 seconds, let it be. It takes more gas to restart it than a minute of idling will use. There are numerous occasions each day when gas could be saved this way: long stoplights, train crossings, freeway congestion, waiting in parking lots or on drive-in bank lines, to name a few. You probably can think of many others.

On average, an engine will use about a gallon of gas for every 1-2 hours it idles. Although this can vary from car to car, on average for every two minutes you idle, you could have driven about one mile. By using the above method you will accumulate good gas savings in short time. Any time the engine is running and the car isn't moving, you're wasting gas. Remember the rule: Under 60 seconds, let it idle; over 60 seconds, turn it off.

Department of Energy tests have demonstrated that restarting an engine "within 8−10 minutes [after it has been turned off] causes little engine wear." Minimizing idle time saves fuel and cuts down on engine wear to boot. Keep in mind that idling is one of the most severe modes of engine operation. It not only uses gasoline, but it also creates engine wear and breeds contaminants in the oil.

After arriving at your destination, turn the engine off immediately! There is one exception to this: When stopping after a continuous high-speed run, it is better to let the engine idle for a minute or so before turning it off. This helps eliminate hot spots and relieves hot fuel vapors that could cause vapor lock and hard starting afterwards.

Keep in mind that, on cars with catalytic converters, long periods of idling can also cause the converter to overheat and damage it. That's another reason to turn it off if a long idle period is anticipated.

SHIFTING GEARS

All owner's manuals have a section on the most efficient road or engine speeds at which to manually shift gears. Consult your car's owner's manual for the most economical shift ranges for your vehicle. See FIG. 1-7 for a typical owner's manual suggested shift range chart.

If you're driving a car with a standard transmission, get into high gear as quickly as possible. On level roads, you should be in high gear before the car reaches 20 mph. Low-range gears use much more gas than their higher counterparts. For example: At 20 mph, second gear uses as much as 15−20 percent more gas than high gear. First gear uses 30−50 percent more! Help your automatic get into the high range faster by easing off the gas just as the transmission reaches its shift point. Getting into high gear fast is one of the economy-run driver's must

UPSHIFTS WHEN ACCELERATING		
Shift From	Recommended Shift Speed	Maximum Shift Speed
First to Second	15 mph (24 km/h)	25 mph (40 km/h)
Second to Third	24 mph (34 km/h)	45 mph (72 km/h)
Third to Fourth	34 mph (55 km/h)	55 mph (88 km/h)
Fourth to Fifth	44 mph (71 km/h)	—

MAXIMUM DOWNSHIFTING SPEEDS	
Shift From	At or Below This Speed
Fifth to Fourth	—
Fourth to Third	55 mph (88 km/h)
Third to Second	35 mph (56 km/h)
Second to First	15 mph (24 km/h)

I-7 Typical shift speeds chart.

techniques. You can use it also, and enjoy the additional miles per tankful it provides.

Skip a gear when you can. Shift directly from low to high if conditions permit. Level or downhill starts are good times to skip second gear (or even to start out in second gear, skipping first) and go directly to third. There are many opportunities, particularly with 4- or 5-speed transmissions, to avoid one or even two gears when shifting. You spend less time in the low, gas-consuming gears and more time in the economical higher ones.

Never shift to a lower gear for the sole purpose of slowing down the car. Except in very hilly country, the brakes will do just fine and you won't have to pay the lower-gear mileage penalty.

Lower gears definitely consume much, much more gas than the higher ones; however, if the engine begins to lug and strain, shift to a lower gear fast, as it will be more economical. The engine uses more gas and works harder when it lugs in a higher gear. A quick glance at the accelerator pedal will verify this. In high gear, with the engine laboring, the accelerator will be nearly floored and the throttle wide open, permitting full flow of gas. By shifting to a lower gear, you get smoother engine response with the accelerator depressed only a fraction of what it was previously. Sure, you are in a lower gear, but you are saving gas. Don't be afraid to shift when the engine dictates it. Of course, with an automatic transmission, the up-down shift cycle is taken care of automatically.

To make starting your manual-transmission car a little easier, place the shift lever in neutral and depress the clutch before cranking the starter. On some newer cars, you must do this to engage a safety switch before the engine can be started. When the clutch is pushed in, the starter has an easier job because it does not have to turn the clutch and gearbox gears. This practice is a safety measure, too, ensuring you will never accidentally crank the starter while the car is in gear.

You can help your automatic transmission by easing up slightly on the gas pedal just before you feel the transmission begin its shift. Easing the accelerator hastens and smooths the shift because of the increased engine vacuum created—and it saves gas, too. With a little practice, you will be able to feel the transmission start its shift and then you can ease up on the gas accordingly. If you have a vacuum gauge, notice how the needle swings into the economy range as you practice this trick.

Cars equipped with a dual-range transmission offer a choice of D1 or D2 driving. Always use the D1, or higher range, for maximum economy. D2 should only be used on the rarest occasions, as decelerating down a very steep hill where the transmission can help brake the car. At all other times use D1.

Many newer cars have three- or four-speed automatic overdrive transmissions that give the driver a choice of OD, D, 1 and 2. For maximum economy always keep the car in the OD position. The car will automatically shift into overdrive at a preset speed and save gas. Most modern transmissions shift into overdrive at speeds somewhere between 45 mph and 52 mph.

Keep the vehicle in overdrive range unless you are moving at a speed where the transmission is continually shifting back and forth between Drive and Overdrive. In this case, it is better to shift to the lower or "Drive" position to eliminate the back-and-forth shifting and wear on the transmission. As soon as the car can be driven steadily at one speed, you should then shift back into "Overdrive."

Use passing gear for emergency situations only.

Jack-rabbit starts are murder on gas mileage. One or two fast starts with the accelerator floored will nullify gains made elsewhere. Take it easy when you start out and push the accelerator down slowly. This is another must for top economy. Department of Transportation tests have proved that jumpy starts and fast getaways can burn over 50 percent more gasoline than normal acceleration.

To avoid hasty starts, envision an egg between your foot and the accelerator pedal. Start out by pushing down in such a manner that the egg won't break. Another good trick is to pretend there is an apple on the front of the hood. Pull away in such a manner that the apple won't roll off. Use these mental tricks, and they will soon form a good economy habit. Remember: no gas-saving device or practice can take the place of a light foot on the accelerator. It's the number-one consideration for top mileage. The lighter the foot, the better the mileage. You must obey this rule if you want top economy.

Some newer cars are plagued by "rotten egg" odor. The catalytic converter treats the unburned portions of the gas/air mixture and if the gasoline has sulfur in it, the results coming out the tailpipe smell like high-school chemistry class.

Some of the ways to help reduce or even eliminate the odor are also economy driving and maintenance techniques. Here are a few of the things to check:

- Be certain the engine timing is set correctly.
- Change brands of gasoline.
- Ask if a change in the air/fuel ratio will help.

- Sometimes recalibrating the engine's computer can help increase gas mileage and decrease rotten-egg odor.
- Go easy on full-throttle accelerations. The gas you waste by putting the pedal to the metal intensifies the odor when raw, unburned gasoline comes in contact with the catalytic converter.

The levelest route with the least number of turns, the least amount of traffic, and the least number of required stops is the most economical.

UPHILL/DOWNHILL

Any time you are going down a grade, take full advantage of the fact that gravity is working in your favor and ease up on the accelerator. Let the car coast of its own volition, even if it means slowing down a bit. Downhill driving affords an excellent opportunity to bolster your gas mileage. A car going up a 4-percent grade at 50 mph gets about 9 mpg. That same car going down a 4-percent grade at 40 mph will get an astounding 35 mpg—a difference of 26 mpg. This is more than enough to compensate for the gas used going up the hill. Use even the slightest downgrade to your advantage. Apply as little accelerator as necessary to keep the car moving at a reasonable speed, and let gravity do the rest.

Know how to drive up a hill. When approaching a hill, try to get a running start at it. Build up as much momentum on the approach to the hill as is safely and legally possible. As the car begins to move up the hill, let the forward momentum carry it and use only as much accelerator as necessary. When you near the top of the hill, ease off the gas and let the car coast up and over the crest. Ideally, velocity should be maximum at the bottom and minimum at the top. On very long upgrades, keep even accelerator pressure, again easing off as you reach the top.

Driving up hills is unavoidable and costly. A 4-percent upgrade can reduce car gas economy by 40 percent, so minimize the effect by using the methods discussed here. Figure 1-8 com-

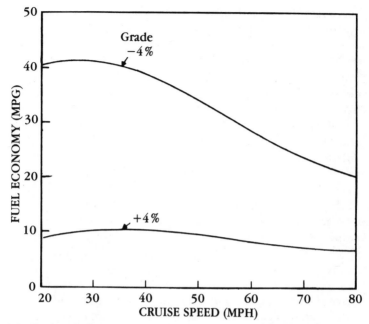

1-8 Estimated effect of grades on fuel economy for a typical, standard-sized vehicle.

pares the estimated effect of different grades on vehicles traveling at various speeds.

Drive into curves properly. Apply brakes and slow down before reaching the curve, then gently accelerate into and through it. In this way, you are in complete control of the auto and get through the curve in the most efficient and safest way.

It should be obvious that if you can avoid driving during rush hour, you'll save gasoline.

Reroute around time- and gas-consuming construction detours. If your day-to-day work route is being repaired, don't keep going back over it. Find an alternate, smoother route with less stops, and use it until road repairs are complete.

Don't tailgate! Much gas is wasted by driving too close to the car ahead. You must constantly brake and accelerate, brake and accelerate. By tailgating, you let the car in front dictate how you drive—and that is foolish economy. Stay a safe and reasonable distance behind the other car, and you won't have to play the slow-and-go game. It's murder on mileage. Keep smooth, even accelerator pressure and pace yourself so that you are not continually slowing and stopping in cadence with the car in front but are keeping an air cushion between the two of you.

Keeping a two- to three-second buffer zone between you and the car in front of you (when possible) allows you space and time enough to react wisely to situations that develop and to make the best use of the momentum your car has built up.

Once you have reached your desired speed, whether in the city or on the highway, keep it constant. You use less gas by going at a steady speed than by intermittently slowing and accelerating. On the highway, varying speed by only 5 mph can reduce economy by as much as 1.3 mpg.

Drive ahead—you see traffic situations developing and are better prepared to cope with them because you are more aware of traffic around your car. You avoid needless slowing and stopping, and conserve critical momentum. Driving ahead can be equated with safety driving as well as economy driving.

Instructors at the Nevada test site for the Department of Energy's Driver Energy Conservation Awareness Training (DECAT) course advise their students to extend their vision about 10 to 12 seconds down the road. This gives plenty of time to plan for various economic modes of driving.

Use your rear-view and side-view mirrors often. Remember to drive behind as well as ahead, and you'll have an advantage in the mileage game. Frequent glances in the mirrors will warn you of situations developing *behind* your car, and you can adjust your driving accordingly. Besides being an obvious safety habit, using the mirrors frequently lets you avoid many turns and stops, saving momentum and gasoline.

Never rev the engine before shutting it off. Too many drivers have this habit. They erroneously think this extra shot will circulate oil for better protection when the engine is off. Actually, the opposite is true. A surge of raw gas floods the cylinders, doesn't have time to be ignited, dilutes the oil, and washes away vital cylinder coatings.

If you are low on gas and feel the engine cough and sputter (hopefully it won't happen to you), pull off the road and turn off the car immediately. This keeps some fuel in the lines and in the carburetor and makes for much easier starting once fuel is replenished. You will avoid long periods of starter grinding, and you won't waste gas with frantic accelerator pumping.

If you do mostly city driving (short trips and stop-and-go driving that never really gives the engine a chance to heat up), take your car out on the highway occasionally and give it a good run at the speed limit. Of course, you can do this as part of a planned trip so you won't waste gas. This will help boil off any contaminants in the oil and keep internal engine parts, including spark plugs, clean and deposit-free.

Almost forgotten these days is one of the most helpful built-in gas savers on your car—the emergency brake! Today's driver practically forgets it's there, and in doing so passes up many gas-saving opportunities. Using the emergency brake when you can saves wear on the clutch or transmission, and conserves fuel to boot. Don't ride the clutch to keep your car at a standstill on hills but use the emergency brake instead. The same applies to cars with automatic transmissions. Holding the car on a hill by applying more gas only wastes fuel and makes the transmission work when it should be resting. Use the emergency brake and let the engine and transmission turn at normal speed.

Don't drive a car with one foot resting on the brake pedal. This is especially important on cars with power brakes, where the slightest pressure will partially engage the brakes. Don't force the car to fight itself—keep your foot off the brake.

In standard-transmission cars, don't rest your foot on the clutch pedal. A small amount of pressure can partially disengage the clutch, causing it to slip and reduce drive-train efficiency. Keep your left foot on the floor.

Avoid sudden stops. They can cause fuel to slosh out of the carburetor bowl or the gas tank and can flood or stall the engine.

When stopped, take your foot off the accelerator and let the engine idle at normal revolutions per minute (rpm). Too often, drivers have a tendency to rest their foot on the pedal—and waste gas as a result. Make it a habit: When stopped, foot off.

Before you start out, make certain the parking brake is fully released. Many drivers let this "silent thief" rob them of mileage they should be getting.

Never pump the accelerator or race the engine while the car is standing still.

When starting a warm engine, whether it is carbureted or fuel-injected, there is no need to push the accelerator pedal down prior to turning the key.

If you have tried everything and still have a problem with engine run-on or dieseling, try leaving the car in gear when turning the key off. This usually does the trick. Be sure to apply the brake pedal while doing this.

Think ECONOMY at all times. This is important. You must motivate yourself to be a successful economy driver. Keep economy and safety foremost in your mind.

Keep a steady hand on the wheel at all times. At first glance, this may not seem important, but it is very crucial if you are to achieve top mileage. Don't let the car wander from side to side. Hold it steady and on a straight course. Any side-to-side movement detracts from forward momentum and can cost you 1-2 mph at highway speeds.

In the old days, when a driver wanted to attract attention, he would turn the ignition off while the car was still moving. At the appropriate time, he'd switch it back on. The result? A loud bang from the exhaust pipe. We weren't doing our cars any favors. Today, engine backfire shouldn't be ignored. The engine is telling you something is wrong. In fact, in one of its recent press releases, the French manufacturer Peugeot said that four consecutive backfires or misfires can destroy a catalytic converter.

Just about every large city has traffic reports during rush hour to advise you of any unusual traffic snarls and suggest ways to avoid them. If you don't already tune into one of the stations that offer these watch reports, do so. They are valuable aids in helping you avoid time- and fuel-consuming traffic jams.

THE EMERGENCY ECONOMY METHOD

It's late at night and you're driving along a deserted stretch of road with the wife and kids. You're very low on gas because the last station you stopped at was closed, but the map shows a small town ahead and you decide to keep going and fill up there. When you reach the town—panic! Nothing there but a few deserted buildings and a railroad crossing. Frantically you look at the map—the next town is 50 miles away! You estimate about a gallon of gas left in the tank, and you know that your car gets around 20 mpg on the highway. A gallon will never take you the 50 miles. Wrong! Not if you know the Emergency Economy Method. That gallon will get you to the next town, and you may even have a little left over! How do you do It? You simply use the coast-accelerate-coast technique of the Mileage Marathon drivers.

Accelerate *slowly* to 20 mph, then quickly turn off the ignition, and shift the car into neutral. Let the car slow to 5-8 mph, start the engine, and repeat the process. That's all there is to it. Repeat this procedure over and over again and you'll be able to double and even triple normal gas mileage. Remember the technique: Slowly accelerate to 20 mph. Turn the engine off and let the car slow to 5-8 mph. Start the engine and repeat the process. Simple, yes—but incredibly effective.

Although this method is obviously impractical for everyday use, it's comforting to know that in the event you are ever caught in an emergency, you can coax many extra miles from your car. I feel it's better to be safe and secure in your car going 5-20 mph than to be stranded and out of gas. If you have to use this method only once in your lifetime, you'll be glad you knew it. **Note**: Do not use this method if your steering wheel locks when the ignition is turned off or when you are driving down steep grades. Also, consult local and state vehicle laws before using the Emergency Method. Coasting, even in emergency situations, might be illegal in your area.

The telephone can save you gas? Definitely! Simply do as the man says and let your fingers do the walking. Use your phone book and call before starting out to a place of business. Make sure that it is open and that the item you want is in stock. Many gas-wasting trips can be avoided by using the phone first.

BREAK IT IN RIGHT

In my book, *Drive It Forever*, I have a chapter called "Break It In Right" that details some of the methods that should be used to break in a new car or a newly rebuilt engine. What does this have to do with fuel economy? Quite a bit. The way a car is broken in has a lasting effect on its subsequent performance and fuel economy. Although there isn't space here to go into a detailed explanation of all the methods used to break in a new car properly, I suggest you pick up a copy of *Drive It Forever*. The difference between breaking a car in correctly and just driving it, when measured in miles per gallon, can be as significant as 5 mpg. That's well worth the time and small effort involved in breaking it in right.

Don't be too anxious to set new EPA gas mileage records with your new car. The fuel economy potential is also breaking in, and it will take time for the car to get its best mileage. Ford says that it takes at least a thousand miles of driving before any consistency will be noted in new-car mpg readings. One thing is for certain, however: During the entire break-in of a new car or rebuilt engine, the fuel economy should be increasing steadily.

The Swedish car manufacturer Saab takes that theory a bit further. According to Saab, "Many new car buyers are disappointed when they trade in a car for a new one of the same model, only to find out it uses more gas." But Saab says this is normal and that the fuel mileage of a car built to tight specifications does not stabilize until it has traveled at least 6,000 miles.

According to Steve Rossi, Director, Public Relations for Saab Cars USA, Inc., "The reason is that the drivetrain's wear surfaces—bearings, for instance—are rough. The roughness is microscopic, of course, and disappears after the initial break-in period. That's enough to make a 10-percent difference in gas mileage."

Using factory-fresh Saabs under simulated urban-driving conditions, Saab found that fuel economy climbed from 19.6 mpg to 21.6 mpg after the first 6,000 miles. After 9,000 miles, the average fuel economy climbed to 21.8 mpg, an 11-percent improvement.

Here's another reason why the fuel economy of a new car may be less than that for the same used-car model: Many car owners often trade in their cars when the new model year begins at the end of summer.

The new-car owners are then comparing apples and oranges; that is, they are comparing the fuel economy of a well-broken-in, older car that has been used in summer with that of a new one running in colder weather. Cold temperatures mean lower fuel economy. When weather is factored into the above scenario, the fuel-economy difference between a new and used car of the same model can be even more than the 10 percent the tests turned up.

Chapter **2**

Cold starts:
That crucial
first minute

If the engine floods while you are trying to start it—strong gas odor and wetness around the carburetor are usual indications—don't continue to pump the accelerator. This wastes gas and makes the car even harder to start. Instead, push the accelerator pedal to the floor and hold it there while turning the starter. This opens the throttle and allows the excess gas to drain out of the carburetor. A minute or two of cranking the starter while holding the gas pedal down will usually get it started. If the engine is equipped with an automatic choke, the butterfly valve must be fully opened while the starter is turned. If the engine is cold, more than likely the valve will be closed. Have someone hold it, or place an object in such a way as to hold the valve open until the engine starts. Make sure the object can't be drawn into the carburetor.

On vehicles equipped with a carburetor, push the accelerator completely to the floor when starting a cold engine. This will activate the automatic choke mechanism and close the carburetor's butterfly valve. One or two pumps should be sufficient to start the car on even the coldest mornings. Excessive pumping wastes gas, could cause flooding, and is probably an indication that the choke isn't adjusted properly.

Check your owner's manual to see how many pumps of the accelerator are suggested for fast efficient cold starts on your car. It's not the same for every vehicle.

The most fuel efficient way to cold- or warm-start a fuel-injected car is to not pump the accelerator pedal before turning the key. Simply turn the key, the engine should start without any encouragement from the gas pedal. If the car balks at starting, only then should the pedal be depressed about half way while cranking the engine.

After a cold engine has been running a few minutes, it is a good idea to tap the accelerator pedal lightly to make sure the high-idle cam has been disengaged. You should notice a considerable drop in engine-idle speed with this easy maneuver.

Years ago it was considered a good practice to let the car idle a minute or two after cold-engine starts. Today, the opposite is true. After you start your car get moving immediately. An engine in proper condition should need no more than 15 seconds of idling before it is ready to go. An engine under road-load conditions will warm faster and lubricate more efficiently than one that is idling. Use slow speeds for the first mile or two, then increase to cruising as the engine gradually warms. Economy improves with distance traveled. Remember to take it easy that first mile or two, because that's when gas consumption is at its peak. Look at FIG. 2-1 to see how fuel economy suffers during the first few miles after a cold-engine start. On a cold day (10 degrees F), a car may never reach its full economy potential in city driving.

Most new car manufacturers now agree with this method of starting a car. Most recommend a minimum of cold idling before moving the car.

Not only does fuel economy suffer immediately after a cold start, but most engine wear occurs then. In fact, McDonnell Douglas engineers have demonstrated that 90−95 percent of all mechanical engine wear occurs during the first 10 seconds of a cold start.

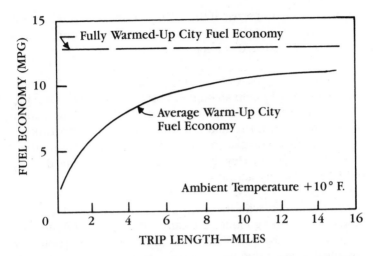

2-1 Average fuel economy in city driving at 10 degrees F, beginning with a cold start.

But long periods of cold idling to warm an engine has proven to be one of the most persistent and difficult-to-break driving habits. It seems logical: Get the engine nice and warm before you put it to work; a warm engine is more efficient and runs better than a cold one.

True, we want a warm engine, but letting it sit and idle is not the best way to warm it. A cold-engine car sitting and idling uses an inordinate amount of fuel because the choke is usually fully engaged during this time.

Most of this fuel is wasted, and nary a drop is used to actually move the car. You get zero mpg during the warm-up. Coupled with this fuel waste is the fact that an idling car does not lubricate itself as effectively as one that is in gear and moving. That speeds up engine wear.

A car that is in gear and moving will warm faster, lubricate better, use less fuel, and realize less wear than one that sits and idles. Contrary to what many drivers believe, idling—especially cold idling—is one of the most severe of all modes of engine operation.

For maximum efficiency *all* moving parts of the car—transmission, wheel bearings, tires, differential—must warm up, and they won't warm if the car is just sitting. Use the above method and take a giant step toward better fuel economy, less car wear, and reduced tailpipe emissions.

The faster an engine warms, the quicker it becomes fuel efficient. One way to help an engine warm faster is not to use the heater until the engine is fully warmed. The heat will remain in the engine and isn't routed into the passenger compartment. Once the engine reaches normal operating temperature, then turn the heater on. Hey, nobody said that economy driving didn't entail some sacrifices.

A car should start with the first turn of the key. Grinding away on the starter isn't good for the car or for its fuel economy, not to mention the battery. When you grind the starter, gasoline is continually being dumped into the engine but doesn't burn. Some of this wasted gasoline finds its way into the upper cylinder area of the engine where it washes away what little lubricant protection the oil pump is providing. Grinding is a no-win proposition and is a sure-fire sign that the engine needs attention.

Putting the pedal to the metal while the engine is cold is foolhardy. The vehicle will use much more gasoline than it would if it were warm. Most of the gasoline of a full-throttle cold-engine acceleration is wasted; it's tossed out the tailpipe or finds its way into the oil where it dilutes and diminishes the oil's effectiveness. No full-throttle accelerations ever is always a good rule; it's even a better one when the engine is cold.

In a test conducted at the Department of Energy's Driver Education Conservation Awareness Training Nevada test site on a 30-degree day, a Dodge Aspen 6-cylinder sedan was started cold and idled for 30 seconds. It was then driven a distance of 2 miles. During the trip, 11 stops were made in an attempt to duplicate the conditions that might occur during a short trip. Although the EPA rating for this car is 21 mpg, during the trip the vehicle averaged just 6 mpg.

Using the same vehicle, the test was extended by 17 miles, duplicating what would happen if a number of short trips are combined. At the end of this extended trip, the mileage had improved to 22 mpg.

In a similar test, a larger V8 sedan was driven under city driving conditions for 2 miles with 12 stops for traffic. The car got only 1 mile per gallon. The same vehicle was then driven in a similar manner for 20 miles and got 17 mpg. The bottom line of both of the above tests: Cold starts and short trips waste fuel and also cause excessive engine wear.

Chapter **3**

Parking techniques

It's much easier and more economical to do difficult maneuvers such as parking when the engine is warm rather than cold. Steering gears, wheel bearings, transmission, and differential also work better when their respective parts are warm. When leaving your car overnight, always park so that leaving in the morning will be easy. Do all backing, turning, and other maneuvering while the engine is warm and using less gas. You'll save a lot with this one technique because you ease the power requirements of the cold engine.

When parking in a lot that has double-row parking (most lots in shopping centers, stadiums, racetracks, and such are arranged in this manner), choose a spot in the forward row. You won't have to back out when you leave, so you're eliminating an unnecessary maneuver.

When you're parked in a street that has a traffic light at the intersection ahead of you, watch the light and gauge your start to coincide with the green. Don't start the engine until the light turns green. Once it changes, pull out into the traffic flow and through the light. You save gas because your engine doesn't idle needlessly, and you don't have to start from a dead stop twice. (You would have had to if you pulled out while the light was red.)

Choose the parking space closest to the intersection if possible. It's easier to get in and out, and you won't waste gas jockeying the car around, as you would in a tight spot.

If no end space is available, try parking so there is plenty of room between you and the car in front. This will make leaving the space a lot easier, and it will save gas.

Before backing into or pulling out of a tight parking place, turn off all accessories. The engine is already laboring overtime turning the power-steering unit, and any added accessory pull foolishly wastes gas. When leaving, wait a few seconds until you are clear and moving before you use accessories—a small sacrifice that pays nice dividends.

In fact, turning off all accessories every time you turn off the ignition will help ease the strain on the engine and save some gasoline when it is restarted.

When you have a full tank of gas, park the car facing downhill if possible to prevent any gas from spilling out of the tank.

Keep your car garaged or under a carport. If this isn't possible, then any type of shelter over or around the car will help. Trees, a wall, the side of a house, all provide some protection from the elements. In cold, windy climates, shelter is a must because moving air can greatly hasten the rate at which an engine cools. Shelter from wind and cold will do more to promote good winter gas mileage than any other single item. Remember: It took gasoline to warm the engine, and anything you can do to conserve the heat will improve mileage.

In hot, dry, or dusty climates, a garaged car is impervious to gasoline evaporation. As much as a quart can be lost on extremely hot, windy days if the car is left outside. Garaging your car also eliminates the problem of dust entering the engine compartment where it can clog the carburetor and the air cleaner. In wet climates, shelter guards against the frustrations of a wet ignition.

Parking in shady areas helps prevent gasoline evaporation that would otherwise occur if you parked in the hot sun. This can be crucial in some southern and western states, where the summertime temperatures can reach 120 degrees and a carburetor bowl of gas can evaporate in a day. Your car will also stay cooler in shady areas, and that means less work for the air conditioner once the engine is started.

There are many instances each day when a driver can save gas by just parking a little sooner and walking a little farther. Why cruise through parking lots time and time again trying to find a spot up front when there are plenty of empty spaces in the back? Take the first available space you see, and don't be afraid to walk the extra 50 yards or so. If everyone practiced just this one item, our nation could conserve millions of gallons of gas each year. Slow stop-and-go driving is the most gas-consuming, so be willing to walk a little—you'll save a lot.

Chapter **4**

Ways to save
at the gas station

Here's how to figure your gas mileage. Jot down the speedometer reading to the nearest mile whenever you fill the gas tank. Let's say it reads 8,510. At your next fill-up, mark down the mileage reading again, and also record the number of gallons of gas purchased. We'll use 8,760 and 10.2 gallons. Be sure to use the fill-up technique described later in this chapter to get accurate gallons-used readings.

Now we subtract our previous speedometer reading (8,510) from the current one (8,760) and get 250 miles traveled. If the speedometer has a correction factor (described below), be sure to add or subtract it from the miles traveled. Let's say our correction factor is +2 for every 100 miles traveled. This gives us 5 extra miles, which we add to the 250 to get 255 as our true mileage figure. Now divide that figure by the number of gallons used: 255/10.2 = 25 mpg. Twenty-five mpg is our gas mileage for that fill-up.

Check mileage with each fill-up. This will give you an accurate running record of how you are doing as an economy driver. As your miles per gallon increase with each fill-up, you'll become encouraged to try more and more economy techniques or additions. Think economy at all times, and let's begin conserving our nation's fuel—and your money, too!

If your speedometer isn't accurate, you may be getting better (or worse) gas mileage than you think. There is no need to take it to a speedometer shop to have it checked because you can do it yourself. Here's an easy, foolproof way.

Take your car out on the interstate or any smooth highway where mileage markers are posted. Note your speedometer reading at the exact time the car reaches one of the markers and write it down, along with the number on the mileage marker. Try to read the odometer to the nearest 1/10 of a mile. For instance, let's say your reading was 21,570.3. Now drive 10 markers (10 true miles) and check the reading again as your car passes the tenth marker. For our purpose, let's use 21,580.1. Subtract the first figure from the second one: 21,580.1 − 21.570.3 = 9.8 miles. You actually traveled 10 true miles, as verified by the mileage markers, but the speedometer recorded only 9.8 miles, so you're getting better mileage than the reading indicates. You must add 2/10 of a mile (9.8 + .2 = 10) for every 10 speedometer miles you travel, or 2 miles for every 100.

Be sure to add or subtract this difference when figuring gas mileage. If you traveled 350 miles on a tank of gas, you would add 2 miles for each 100 traveled—in this case, +7 miles. So the true corrected mileage is 357 miles.

Of course, it can work to your detriment as well. If the odometer records more miles than you actually travel, then the extra mileage must be subtracted from the total miles to find your true mileage.

FILL THE TANK RIGHT

If you use a full-serve gasoline station try to get away from the habit of saying, "Fill it up." Chances are, the attendant will fill it up and then a bit more. It's not uncommon to see them over-fill the tank every time. Although most new cars have provisions to avoid overfilling, older cars don't and spills are a too frequent sight in full-serve stations.

Insist on a slow fill-up if an attendant waits on you. Gas that spills down the side of your car because of an over-anxious attendant still costs you, but you don't get to use it.

How many times have you noticed gas dripping from the tank of the car ahead of you? It's easy to lose a half-gallon of gas by filling the tank to the brim. Every time the starts or stops, gas sloshes out of the filler neck. If the day is hot, the gas expands and forces itself out. Don't top off the tank, but keep the gas level a comfortable margin below the cap.

A good way to fill your tank without spilling and wasting gas is to set the automatic shut-off at the slowest delivery rate. Place the nozzle as far into the filler neck as it will go, and when the pump clicks off, round out the amount to the nearest $1/10$ of a gallon. This leaves the tank a gallon or so from full and avoids gas spills. It also gives you an accurate method of determining the exact number of gallons used when figuring gas mileage. Try to use the same pump and park the car in the same direction to get an even more accurate reading.

We are caught in a Catch-22 situation each time we get gasoline. If we fill the tank, we must lug more weight around until the tank level gets lower. If we don't fill the tank, we are inconvenienced by having to put gas in more frequently. In addition, low fuel invites condensation.

Fill the tank—but don't overfill it—at each stop, and use up most of the fuel before you refill. That way, you average out the weight. The mileage gained by driving with the tank less than half full offsets driving with the tank over half full. Theoretically, you should get a little better mileage for each mile you travel because you are continually using up fuel and thus reducing the weight in the gas tank. And, maybe more important, filling the tank gives you a chance to check your mileage—an absolute must for every economy driver.

Buy your gas in the early morning whenever possible. In the cool of the morning, gas is more dense because the sun hasn't had time to heat and expand it. Gasoline pumps measure volume, and although you receive only the amount the pump indicates, it is more concentrated.

As the day wears on, the gas will expand in your tank and you gain the extra expanded volume. If you bought the same amount of gas in the heat of the day it would have already expanded, and as the evening approached and the gas cooled, you'd lose volume, because gas seeks its cooler, more dense state. Try early-morning or late-evening fill-ups, and profit from the extra gas you receive at no additional cost.

When recording gas mileage, make a note of the brand of gas purchased at each fill-up. If one brand consistently gives better mileage than the others and is competitively priced, it makes good sense to use it. Sometimes brands of gasoline can make a difference, so compare and save.

Keep your gas tank on the full side to prevent gasoline loss from evaporation and condensation. This practice can be especially helpful in the winter when condensation from a near-empty tank could freeze, blocking gas lines and making starting impossible.

If you frequent one of the popular self-service gas stations, don't be embarrassed to lift the hose at the end of gas delivery and wait for all the gas to run out. You can gain an additional half cup by this simple practice. You wouldn't think of leaving an item on the checkout counter of a supermarket, so why do it at a gas station? You need every edge you can get in the battle for better mileage.

Although gas wars are probably a thing of the past, you can still keep your eyes open for low-priced gasoline. Service stations a block or two from freeway off-ramps may offer the same brand of gas 10 cents a gallon cheaper than their freeway counterparts. It pays to drive a bit and look around. It takes but a minute or two of your time and you could end up saving a couple of dollars on a fill-up.

Whenever you take a long trip (one that requires at least one fill-up on the road) and plan to return by the same route, write down the names of the towns and filling stations that offer the lowest gas prices. Then, on your return, take advantage of these gasoline oases by timing fill-ups to coincide with the respective towns. Any traveler knows how gas prices can vary from town to town, so follow this easy method, and pocket the extra savings.

Here's a tip that can save you many gasoline dollars when driving in a foreign country. Be sure—stand there and watch—that the station attendant resets the gas pump meter to zero ($00.00) before putting gas in your car. Literally thousands of dollars have been fleeced from unsuspecting tourists who were charged the amount of the previous sale, plus the amount it took to fill their own tanks.

In the U.S., most pumps are automatically zeroed or must be zeroed by hand before delivery can begin, but in many foreign countries this isn't the case. Get out of the car and make sure the attendant "zeros" the pump before he begins filling your tank. I learned this the hard way many years ago in Mexico, paying for 120 liters of gas when the tank could only hold 80. Some country stations in the U.S. might still have the old pumps, so it's better to get out of the car and make sure if you have doubts.

Caution! Never use leaded fuel in a car meant for unleaded fuel only. Even if you got better mileage, you would ruin the catalytic converter in the process. What's more, it's against the law.

OCTANE—DON'T BUY MORE THAN YOU NEED

Don't pay for octane your car doesn't need. It's money down the drain. Plying your car with $1.70 premium when it will run as well on $1.40 regular is like the guy who takes huge quantities of vitamins each day, only to have his body reject what it doesn't use. The body knows how much of each vitamin it needs and uses only that amount. The same applies to your car, so use the lowest octane that provides good performance. You can save more than $3.00 on each tankful by knowing the exact octane requirements of your car.

If you live in an area where octane selection is limited to regular and premium, here's an easy and economical way to custom-blend the exact octane your car needs. If the car doesn't need premium gasoline but seems to knock or run a little rough on regular, the correct octane lies somewhere between the two. On your next trip to the gas station, custom-grade your own. Fill the tank two-thirds full with regular and then top off the remaining one-third with premium. Experiment with each tankful, adding or subtracting a little premium until you find the exact mix that makes your car run best. Mark down the amount of regular and premium used, and then figure the ratio of the two.

Let's say you bought 10 gallons of regular and 5 gallons of premium; the ratio then is 10 regular to 5 premium, or 2:1. So every time you fill up, use the 2:1 ratio and pay only for the octane your car needs. The lowest octane that gives smooth performance is the best buy.

If your car uses premium gasoline, you can save money by purchasing regular whenever you take a trip over mostly level roads. At highway or cruising speed, engine load is only a fraction of the horsepower available. A car that must use ethyl under city driving conditions may do surprisingly well on regular. Try a tank of regular, or a mix of one-half regular and one-half premium, and you'll be pleasantly rewarded with equal response at less cost.

A car that must use premium gasoline at sea level will probably run just as well on regular gasoline at high elevations (5,000 and above). The reason is that as altitude increases, the car's octane requirements are lessened due to the change in the air/fuel mixture. Thinner air at high elevations means a richer air/fuel ratio, and less octane is needed to fire the carburetor. If you are in the high country awhile, try switching to regular, and save the substantial difference per gallon.

Many newer-model cars that use unleaded gas experience engine knock, rough idling, and run-on (the engine continues to run after the ignition has been turned off). Switching to another brand of unleaded will sometimes cure these problems. Some brands have higher octane ratings than others, and a few octane/numbers higher may be just what the doctor ordered to cure the problem.

Even though two different brands of gasoline might have the same octane number, one might give better performance and fuel economy than the other simply because of the way it was refined and the additive package used in it. Before you switch to a more expensive, higher-octane gasoline, try switching to another brand of gasoline that has the *same* octane. In many cases, you will be pleasantly surprised with better mileage at no extra cost.

WHAT YOU SHOULD KNOW ABOUT OCTANE

What is octane? Simply stated, it is the measure of the gasoline's ability to resist engine knock. What does the (R + M) number on the gas pump mean? It is the Research Octane Number plus the Motor Octane Number, divided by 2. The resultant number is commonly known as the (R + M) anti-knock index. It is the number you have seen posted on all gas pumps and the one with which you should be concerned.

How many times have you heard someone say that he owns such-and-such model car and it runs great on regular gasoline, even though the owner's manual says to use premium? Yet you own the same model car with the same engine and it won't even start, let alone run, on regular gas. Is the first guy just pulling your leg? Probably not. Due to assembly-line variances in engine tolerances, octane requirements for the identical new car can vary by as much as 10 numbers! His car could be doing very well on 86 octane while yours could require a quite higher grade.

Another thing to remember about octane: As your new car accumulates mileage, the octane requirement continues to increase until combustion deposits (carbon) stabilize. Octane requirement usually levels off at about 5 numbers higher than when the car was brand new. This means that an older car will always require a higher octane than when it was new.

Gasolines have improved tremendously since I wrote the original edition of this book in 1978. However, all gasolines aren't—as many believe—the same. Each gasoline is different, and each will leave its unique signature on your vehicle. That signature determines how the engine performs, its fuel economy, how easy it starts, and other operation parameters—even, to some extent, how long it will last.

For today's cars, it is absolutely imperative that you use a gasoline that has a high-detergent content, a proven fuel system cleaner/conditioner, and a deposit-control additive. The fuel-injection systems of modern cars are very critical of the cleanliness of the gasoline used in them. Poor quality, low-detergent fuels will clog fuel injectors, causing driveability problems and reduced fuel economy.

Most major brands of gasoline are "clean," high-detergent fuels and are fortified with fuel-system cleaners and deposit-control additives. Most of them advertise this fact. Just a few brand names that are good choices include Chevron, Texaco, Shell, Mobil, Exxon, Amoco, Sunoco, Union 76, and Phillips 66. There are more, but this list should be a good starting point.

Fuel injectors must be clean if you want top mileage from your car. What's the best way to keep them clean and operating freely? Use a high-detergent major brand of gasoline with a proven fuel-system cleaner and conditioner. It's false economy to use Joe's Freeway Special because it costs 5 cents less per gallon. In a short time, your car will object with clogged fuel injectors.

Don't be penny wise and pound foolish when it comes to gasolines. All gasolines are not the same. Use a quality, high-detergent brand, and you'll get top performance and efficiency from your engine.

Although carbureted vehicles aren't as critical of their fuel as are fuel-injected ones, gasolines without a correct fuel additive package will contribute to deposits in carburetor throttle bodies, carburetor idle circuits, PC valves, and EGR systems. These deposits will negatively affect fuel economy.

The volatility of a gasoline plays a major role in determining how fast the engine will heat. More volatile gasolines permit the engine to heat faster and get better mileage. One test demonstrated the difference in warm-up distances using two different gasolines. With one gasoline, the distance required was 2.7 miles, while with the most volatile gasoline, the distance was reduced to 1.6 miles. That's very significant when you consider that this warm-up period is the time of highest fuel use.

Although you can't determine how volatile a gasoline is by looking at the pump, you can experiment with different brands and choose the one that gives you the easiest starts and the best performance under cold conditions.

Regional custom-blended or climate-blended gasolines contain specific additives that help the gasoline perform under the climatic conditions of the area in which they are sold. In cold climates, they will contain volatility improvers to ease starting and to promote rapid warm-up. These are a good starting point in your search for a rapid warm-up gasoline.

GAS PRICING OWED TO BILLY JOE

Have you ever wondered who sets those gasoline prices you see changing just about every day? I found the answer a number of years ago and reported on it in a syndicated column. Read on, but keep an open mind and tongue in cheek.

Before he passed away a few years ago, Dan Lundberg, publisher of The Lundberg Letter, a highly respected oil-industry newsletter that tracks, among other items, the average price of gasoline across the United States, called me with a request. He wanted to reprint a column I had done on gasoline pricing. The request was unusual because The Lundberg Letter was all business, while the column in question was tongue-in-cheek. But Lundberg loved it and said my explanation of why gasoline prices rise and fall was as good as any he had seen.

The recent and dramatic rise in the price of gasoline at the pump has encouraged me to get the real truth about gasoline pricing out to as many readers as possible. Although at first glance it might seem that the crisis in the Gulf is responsible for those suddenly high prices at the pump, that explanation—much to the relief of Giant Oil, I'm sure—is far from the truth. As you will see, the timing of the pump price gouges and Saddam Hussein are purely coincidental. Here is the real story behind the latest dramatic price increases at the pump.

The following section has been reprinted by permission of the New York Times Syndication Sales Corporation from Bob Sikorsky's "Drive It Forever" column.

I recently spotted a clerk at a convenience mart changing gas-price signs and stopped to ask her how they determined when and how much to raise or lower their prices.

"It's simple," she said. "We just take our cue from the Exxon station across the street. If they raise or lower prices, so do we. All our stores' gas prices are keyed to the prices of other nearby gas stations."

Intrigued by her answer, I send this column's top investigative reporter—me—to find out more about this unique pricing structure.

I went to the Exxon station and asked how they determine the price of their gasoline. I learned that they, too, were keyed to another station—this time a Texaco just down the road. The Texaco station, it turned out, got its cue from a Shell station a few blocks away. This domino-like structure was prevalent everywhere, each station raising or lowering its prices when its cue station did. Determined to find out where the chain ended, I pushed on, feeling like an automotive Jack Anderson.

My quest led me all over town and eventually took me out of the state, where I discovered many such similar pricing chains spiraling out over the whole nation like the spokes of a wheel. A few months and a few thousand gas stations later, the answer began to form as I followed these spokes back in the direction I had come.

I finally found the hub of the price-structuring wheel in a most unlikely place—on a dusty back street in Gila Bend, Arizona. There it was, in all its dilapidated glory, Billy Joe's Gas-n-Go Market! I couldn't believe it. What I had discovered was enough to put the lid on any and all international oil-conspiracy theories for good. Here, at this old, wooden grocery store and gas station, the nation received its cues on when to raise or lower gasoline prices.

I watched as cars from every oil company cruised the street, the occupants' eyes locked on Billy Joe's gas-price sign. I asked questions. Rumor has it that the late Dan Lundberg, whose Lundberg Letter is considered the bible of gasoline-price predictions, has an apartment across the street. A café nearby was filled with white-robed, bearded Middle Eastern-ers, oil sheiks in a different desert. Was that Mo Udall sipping coffee with another House member? I jumped out of the way as a Department of Energy car whizzed by, the driver jotting notes on a pad. It was too much; Billy Joe's Gas-n-Go, the hub of the pricing wheel, the place where every oil company and producer got its cue to raise or lower prices.

But what made Billy Joe raise and lower his prices? I had come this far and wasn't about to quit. I just had to find out.

Late one afternoon, an opportunity presented itself as I spotted Billy Joe himself rocking away on the rickety front porch of his establishment. Billy Joe, a good ol' boy by way of Apalachicola, invited me to set a spell. I accepted, buying a six-pack at inflated prices so we could better ward off the dust cloud generated by all the passing cars. As we chatted and the beer began to loosen up the good ol' boy, I asked him what made him raise and lower his gasoline prices. Billy Joe, unaware that his place was the origination point of interna-tional oil- and gasoline-price structuring, thrust forward in his rocker and spewed a stream of Red Man into the dusty parking lot.

"'Tain't no secret to it," he explained laconically. "Ya see, I jest loves poker—play two, maybe three times a week. If I lose a little, I raise my prices a little to make up for my loss. If I lose a lot, I raise my prices higher and hold 'em there longer. If I win, which ain't too often," he chuckled, "I lower 'em to celebrate."

I couldn't believe my ears. So much for OPEC and those much-heralded pricing meetings. Forget the Giant Oil greed and conspiracy theory. Ignore Saddam Hussein. Throw away supply and demand. The answer was here at Billy Joe's Gas-n-Go.

So the next time you notice gas prices creeping upward— as indeed you have been noticing lately—don't look to the Department of Energy for an explanation and don't bury your-self in the financial section of the newspaper or call some pro-fessor of economics looking for an answer. Just pray that Billy Joe starts getting some good cards.

Chapter **5**

Gas-saving items
for your car

The best way to learn economy driving is to install a vacuum gauge in your car and let it be your teacher. The gauge is clearly marked to indicate when you are driving economically and when you are wasting gas. It monitors engine vacuum, which varies with the amount of pressure you apply to the accelerator pedal. If you watch the gauge, you can't help but improve your gas mileage. A vacuum gauge can be purchased for as little as $10, and it is simple to install. You also get a valuable added bonus because a vacuum gauge can detect a whole range of engine ailments before they become serious. By correcting these immediately, you save yourself costly bills later. If you buy only one add-on piece of equipment for your car, make it a vacuum gauge (FIG. 5-1).

In lieu of a vacuum gauge, there are vacuum- or engine-rpm-actuated lights that warn the driver when he is wasting fuel. Although not as accurate and telling as a vacuum gauge, these lights are helpful. Some newer cars have these lights, called *shift lights*, built in as standard equipment.

A tachometer can be a helpful instrument for improving driving technique and gas mileage. The tach will indicate engine

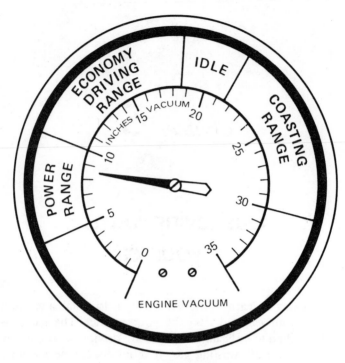

5-1   A typical vacuum gauge.

rpm, giving the driver the correct and most efficient point at which to shift gears. You avoid lingering in the gas-consuming lower gears and eliminate costly and damaging engine lug by shifting at the proper times.

A fuel-consumption meter or *fuel-flow meter* is a device that measures the exact miles per gallon your car is delivering at any time or for any given distance. It is accurate to $1/100$ of a gallon. It is a must for the true gas-mileage enthusiast because you can actually see what mileage you are getting at any particular moment. The only drawback it has is price. Cost per unit is in the $60 to $80 range. This meter and a vacuum gauge make for probably the best combination of gas-saving devices available today.

Water vapor, or water-alcohol vapor injectors, can be helpful on some cars. Ever notice how your car seems to run better on rainy days? Moisture in the air creates a more even-burning fuel mixture, and the car seems to respond with smoother performance. Vapor injectors use the same principle: Add moisture, and get increased volatility from the air/fuel mixture. During World War II the U.S. Government experimented with water-vapor injection, and it is still being considered today as a method of conserving our rapidly dwindling fuel supply.

A water-alcohol mixture, when injected into an internal combustion engine, will improve performance and increase gas mileage. However one must consider whether the price of the alcohol is offset by the increased gasoline savings. Many brands of water-alcohol injectors are now on the market, at prices from $20 to $40. They all employ the same basic principle: Additional outside air is drawn through a container of water-alcohol, causing the mixture to bubble. The mist from the bursting bubbles is then drawn into the intake manifold at some point below the carburetor and sucked into the engine. As the water-alcohol mist combines with the normal carburetor mixture, it produces a better-burning, cleaner fuel, enhancing the car's pep and fuel economy. It may be worthwhile to experiment with vapor injection on your car.

Consider the price of the injector unit and the cost of a container of methyl alcohol (a half gallon usually lasts from 4,000 to 7,000 miles). Weigh these figures against any increase in miles per gallon you get, and then decide for yourself if that increase will eventually offset the money spent for the unit and the cost of additional alcohol refills.

Caution: Methyl alcohol, or *methanol*, has been shown to be corrosive in the engines of some vehicles. Check with the manufacturer of your vehicle before installing an alcohol injection unit. On some newer vehicles, the injection unit could adversely affect the operation of the engine's sensors and computer. The installation of such units could be considered as emissions-system tampering.

Some cars come equipped with a magnetic oil-drain plug that attracts loose metal particles suspended in the oil. If your current auto doesn't have one (chances are it doesn't), you can purchase one at many auto accessory stores. Or, if you like, simply attach a small magnet to the inside of your present drain plug. By attracting oil-suspended metal particles, the magnetic plug reduces abrasion between moving parts, thereby lessening internal friction, preventing engine wear, and promoting economy. Be sure to clean the particles off the plug at each oil change—you'll be surprised at how many the magnet attracts.

Spending a few dollars on a locking gas cap could save you from being siphoned to the tune of a $25 tank of gas. It also protects you against pranksters throwing dirt or foreign objects into the gas tank. If you do not have a locking gas cap on your car, purchase one as soon as possible. Use the gas you paid for. Don't let your car be an open invitation to gas thieves.

Some cars feature as standard equipment a speedometer or tachometer buzzer that can be set to go off at a preselected speed or engine rpm. During monotonous highway cruising, drivers have a tendency to speed up as the accelerator foot becomes heavier. The buzzer is an excellent way to prevent this from happening because it alerts the driver and warns him to reduce speed. It should be standard equipment on all cars. It's a gas saver and an excellent safety device. Some auto supply stores have the buzzers at modest cost, but installation may require some skill.

On older cars, installation of an automatic-choke conversion kit is a good economy move. Costing only a few dollars and relatively simple to install, it provides dash-mounted fingertip control of the carburetor choke valve. With the dash control, you use only as much choke as is needed to start the car. The problematic automatic choke with all its cleaning, adjusting, and malfunctioning is eliminated. It takes but a few cold-engine starts to find the proper hand-choke setting. Engine feel will let you close the choke much sooner than with an automatic, lean the mixture according to engine needs, and operate at peak cold-engine efficiency. This is very critical if you are to obtain good mileage. A cold engine can use twice the amount of gas as one already warmed, so anything that lessens gasoline consumption while the engine is cold will result in double savings.

You might also consider investing in a hand throttle, along with the hand choke, for your older vehicle. A hand throttle lets you vary the idle speed from the driver's seat and is helpful during warm-ups where a hand choke is used. In essence, the hand throttle replaces the high-idle cam on automatic chokes and is relatively trouble-free.

Fuel-pump pressure regulators are options available at a modest price. Their purpose, as the name implies, is to provide correct, even-pressured fuel flow to the carburetor. They can have some value if your car tends toward overrich fuel mixtures and all conventional attempts to correct it have failed.

Cruise control can be a valuable gas-saving addition to any car. When driving at highway speeds, you can set the control to your desired speed and it automatically takes over the accelerator pedal, gently and efficiently applying or easing pressure as conditions dictate. Steady and even speed is maintained with minimum throttle, thus conserving fuel. (It is not wise, however, to use cruise control in mountainous country.) Add-on electric or mechanical cruise-control units can be purchased for as little as $30.

"A manual transmission is generally more fuel-efficient than an automatic. In fact, a four-speed manual transmission can provide a fuel savings of five percent over a three-speed automatic. Since incorrect use of a manual transmission can waste gas, choose a transmission that matches your needs and experience. Transmissions featuring an overdrive gear can improve a vehicle's fuel economy by as much as nine percent for an automatic, and three percent for a manual transmission."

U.S. Department of Transportion

Many new cars offer overdrive units as standard equipment or as an available option. If you are buying a new car and plan to keep it awhile, by all means purchase one with overdrive. It will cut down dramatically on gas consumption and reduce engine wear by allowing the engine to work only a fraction of what it normally would at higher speeds. Overdrive is an excellent investment that will pay for itself in a short time by improving highway fuel economy up to 25 percent.

Instead of overdrive, most manual-shift new cars come equipped with a 5-speed transmission, the fifth or highest gear acting much the same way as an overdrive. This is a marvelous built-in gas saver, especially for highway driving. Consider a 5-speed transmission when shopping for a car, and enjoy the additional gas-saving benefits it provides.

When having a new muffler installed, make sure it is the proper fit for the year, make, and model car you drive. A muffler that creates excessive back pressure, not allowing free passage of exhaust fumes, is disastrous to gas mileage. It could also cause severe power loss and eventually damage the engine. Low back-pressure mufflers are great for gas mileage.

Install a scavenger tip(s) on the exhaust pipe(s). This is a tip in which the opening points downward instead of straight out. The advantage it gives is that the onrushing air has a tendency to suck the exhaust from the pipe, easing engine back pressure and allowing for a more rapid discharge of exhaust gases, which has a positive effect on mileage.

In my book, *Drive It Forever*, I devote over two full pages to the merits of engine heaters and how they can reduce engine wear and improve gas mileage. There are a number of kinds of engine heaters; some warm the motor oil, while others warm the radiator coolant. Either type can work wonders for your fuel economy.

Plug in an engine heater when you retire, and the motor will be prewarmed in the morning. That means the vehicle doesn't have to go through the very fuel-inefficient and wear-intensive cold-start period discussed in Chapter 2.

Besides being the time of accelerated engine wear and very poor fuel economy, the cold start is also the time of highest emissions. If an engine heater is not a part of your fuel-economy equation, I recommend you purchase one. The heater will soon pay for itself in fuel savings alone; the many extra miles of car life and reduced emissions are a significant bonus.

You can turn your cold starts warm for as little as $10 – $30 for an engine block heater. Oil dipstick warmers and magnetic, clamp-on warmers can be purchased for as little as five dollars.

If you don't have access to one of the above warmers, place old blankets or rugs over the engine. They will help retain some of the heat. Some cover is always better than none. Pre-shaped, insulated blankets, made expressly for this purpose, can be purchased through many accessory catalogs or at automotive specialty stores.

Another heat-retaining method is to place a 100-watt household bulb under your car's hood, near the battery. This will help prevent cold from lessening the battery's cranking power. A strong battery means easier cold-engine starts with less wasted gas. The bulb will also guard against fuel line freeze-up and help keep the engine oil a bit warmer.

A heater shut-off valve installed in the heater hose leading into the passenger compartment can save gas in the summer. This valve shuts off the flow of hot coolant into the passenger cabin and helps keep the interior cooler. That means less work for the air conditioner because no hot coolant is flowing near the A/C's evaporator unit (located under the dash). The bottom line is that less gasoline is needed to run the A/C compressor to keep the car cool.

Automatic garage-door openers can save gasoline. Because you can remotely open the door you can drive directly in or out of the garage without waiting. Otherwise, the car would have to sit and idle while you open or close the door.

Don't have a remote garage door opener? Then let whoever is riding with you do the chores. If you are alone, ask whoever is at home to close the garage door after you leave.

A dash-mounted timing selector is a low-cost add-on device that can improve older, pre-computer vehicle performance and increase gas mileage if used properly. It allows the driver to advance or retard distributor spark according to driving conditions and engine load. Hot and cold engines, idling, acceleration, cruising, city driving, hill driving, and driving with additional weight all (ideally) demand different degrees of timing for maximum economy. Factory distributors are preset, and the ignition timing they produce is, at best, a compromise between highway and city driving needs. With a timing selector, you can choose the exact degree of timing needed for each type of driving, and get top mileage for your efforts.

Ford Motor Company says that a hotter degree thermostat will provide better engine economy than one of a lower heat range. Their reasoning is correct, too, because a warmer engine always uses considerably less fuel than a cold one. The lower (160-170°F) degree thermostats are mostly used when the cooling system contains a pure alcohol-type antifreeze. Because alcohol has a relatively low boiling point, it is necessary to run the engine on the cool side to prevent boil-overs and loss of coolant. It's rare to find the old alcohol-type antifreeze anymore, but if by chance you are still using it, switch to the permanent summer-winter type, and install a hotter thermostat to obtain better gas mileage.

If you're buying a new car, order one with a lower, more economical rear-axle ratio. It doesn't cost much more and is a practical investment that will pay for itself in future gas savings. With a lower axle ratio, the rear wheels revolve more times for the same amount of engine work as compared with a standard axle ratio, and you can go farther with the same amount of gasoline. In government-sponsored tests of possible fuel-saving methods, a lower rear-axle ratio was recommended as a feasible, economical, and readily available fuel-conserving modification.

The U.S. Department of Transportation defines *axle ratio* as "the ratio of the revolutions of the driveshaft (transaxle for front wheel drive) to the revolutions of the wheel." It goes on to say that "generally, a low ratio, such as 2.53:1 means *less engine wear and better fuel economy* [emphasis mine] than a higher ratio, such as 3.55:1. The lower axle ratio will result in slower acceleration, but it will pay off in increased fuel economy in highway driving. In general, a 10-percent decrease in the axle ratio can improve fuel economy by 4 percent."

A prospective new-car buyer should be aware that axle ratio is simply a numerical designation for the various rear-axle gear combinations that can be built into the car at the factory. It is not something you can go out and buy at the local auto parts store.

Each car comes with only one axle ratio, either high, low, or in between. For instance, a car that has 3.55:1 axle ratio requires the driveshaft to turn 3.55 times in order to get the wheels to turn one complete revolution. With a 2.53:1 ratio, the same car would need only 2.53 revolutions of the driveshaft to accomplish one revolution of the wheels. For each ratio, a different set of gears is used in the axle.

Many car shoppers are not aware that they have such a choice. When available, a low axle ratio is a smart pick.

Extra-high-resistance ignition wires require more voltage to fire the spark plugs efficiently. If you have a high-intensity coil, the coil can deliver enough voltage to overcome the added resistance of the wires. However, high-resistance wiring, in combination with a stock or weak coil, can cause a drop in engine performance and economy. For best mileage, stay with the less resistant, straight-through metal core wires with silicone-rubber insulation; they're more heat-resistant and will last longer, too.

A weak ignition coil won't allow maximum voltage to reach the spark plugs. Low or erratic voltage at time of detonation results in incomplete combustion because the spark supplied by the coil is not hot enough to fire the entire charge. The unburned portion goes out the exhaust or is deposited on the cylinder walls. Either way, it costs you money.

If coil replacement becomes necessary, invest in one of the new high-voltage kinds. They can be purchased for approximately the same money as a factory-type coil and will ensure peak voltage to the plugs under all load conditions. You enjoy better combustion, less engine carbon buildup, cleaner, longer-lasting spark plugs, and increased economy.

A capacitive discharge or an electronic ignition will improve mileage and extend spark plug and valve life. As a rule, one of these improved ignitions is expensive and will pay for itself only if the car is driven considerably or kept for a few years. There are many of these on the market, so be selective and choose a brand that has a good reputation.

On cold winter days you've probably seen big trucks (and cars, too) rolling down the highway with the radiator partially covered. By blocking some of the on-rushing cold air that normally passes through the radiator, a warmer and more efficient engine results. Cold morning warm-ups, the most critical period of gas consumption, will also be hastened. A small piece of cardboard or vinyl cloth placed so that it covers a portion of the front of the radiator will do the job nicely. Be sure to remove it as the weather gets warmer.

If the time ever comes when you have to replace the carburetor, consider buying a smaller one for better economy rather than an original-equipment model. It can be easily adapted to your present car, and chances are it will cost less than the original. Sometimes the fuel savings gained by switching to a smaller carburetor can be dramatic.

If you are mechanically inclined and won't mind rebuilding the carburetor, mileage-economy kits can be purchased for many models that will do the same job as the exhaust-gas analyzer method described in Chapter 6. Leaner jets, metering rods with longer economy steps, and redesigned accelerator pumps are supplied, along with the basic elements of standard rebuilding kits. Mileage improvements can run as high as 4 mpg, well worth investment in the kit.

A wire-mesh screen (about 8 grids per inch) inserted between the carburetor flange gasket and the intake manifold will help break up the air/fuel charge and produce a more homogeneous mixture for combustion. Also, on rapid acceleration, excessive fuel and moisture tend to cling to the screen momentarily, instead of being drawn instantly into the intake manifold. This helps stop raw gas from flowing into the cylinders and diluting the engine oil. Make sure that the mesh is sealed tight and that no air leaks are present. Some new cars are equipped with a heated grid that helps vaporize fuel under cold start conditions.

A homemade ram air charger, or an inexpensive store-bought unit, will add extra miles to each tank of gas. If you do considerable highway driving, it can be especially beneficial. Dense outside air is force-fed through flexible ductwork directly to the air cleaner. As the car's speed increases, more air is supplied to the engine. Most engines are air-starved at higher speeds, and ram air is one way to supply the additional quantities needed for more economical performance.

To make a ram air charger yourself, buy a length of flexible duct hose similar to the type used under the dash for your car's heater blower. Fix one end so the opening will scoop up onrushing air—to either side of the radiator, behind the front grill, is a good spot. Then cut off a section of the air cleaner snout so its diameter approximates that of the duct, and attach the duct to the enlarged opening. If you wish, cut a circular hole in the side of the air cleaner, and attach the hose at that point using small sheet metal screws.

If for any reason the cylinder heads must be removed from the engine (in case of a valve job or other such repair), you might consider having them milled a fraction, or install an extra-thin head gasket when they are replaced. Hot-rodders have used this trick for many years as a basic way to increase power and economy by raising the engine's compression ratio. The simplest and least expensive method is to use the ultra-thin head gasket (you have to have one, anyway) instead of one of standard thickness.

Race-car mechanics know of the additional power and performance gains attributed to customized or high-performance camshafts. On the other hand, economy buffs know that cams can be customized for economy, and they are willing to accept some power loss to gain extra mpg. However, installing an economy cam is a costly procedure and should be considered only if a major engine overhaul becomes necessary.

A dual-intake manifold, popular with hot-rodders and performance-conscious drivers for years, will increase gas mileage by balancing the fuel mixture to each cylinder—in effect, making it possible to use each drop of gas more efficiently.

Designers of high-performance racing cars are well aware that relatively small changes in body shape can result in fuel savings of up to 12 percent at high racing speeds. These improvements in vehicle design will also provide savings at passenger car speeds. Adding a spoiler to your car is one simple, effective way of altering the vehicle's shape and reducing aerodynamic drag. A spoiler is nothing more than a piece of flat sheet metal or fiber glass that is fixed underneath the car, immediately behind the front bumper. It is fastened so that it covers up the array of protruding engine and front-end parts and negates their drag by smoothing out the underneath profile of the car. Mileage gains of 1-3 mpg are not uncommon when a spoiler is installed. Indeed, it is well worth the time and the few dollars it takes to make one.

Since the original edition of this book appeared in 1978, vehicles have become more aerodynamically efficient. A popular aero item is the spoiler. While it's certainly possible to make a crude homemade spoiler to place underneath the front area of a car, a number of aftermarket companies now offer spoilers as add-ons.

Chin spoilers and air dams reroute the oncoming air more efficiently and direct it away from the underneath of a car. Many car enthusiasts buy these for looks and aren't concerned about the improvement they may get in fuel economy. At higher road speeds, spoilers can reduce wind drag and help improve the vehicle's fuel efficiency. However, their ''payback'' period may be a long one. It's up to the individual car owner to decide if a spoiler is worth the investment. Resale value of the car—enhanced by the addition of a spoiler—appearance value, driver satisfaction, and increased fuel economy when added together may make a front-end spoiler or air dam a good investment.

Whale tail or other types of rear-deck lid spoilers can be added to the top-rear portion of the trunk to give the car an appearance boost and help increase fuel economy at highway speeds.

Kamei spoilers are one of the few items tested or evaluated by the Environmental Protection Agency that showed "a statistically significant improvement in fuel economy without an increase in exhaust emissions" when retrofitted to a vehicle. (See Chapter 11.)

However, the EPA goes on to note that the "cost-effectiveness must be determined by the consumer for his particular application." Front or rear spoilers can improve fuel economy. It's up to you to decide whether the investment is worth it.

If you drive a van or large truck, or if you do a considerable amount of trailer or camper pulling, consider investing in a wind deflector. These easily attached devices cut wind resistance considerably and can add many extra miles to each tank of gas.

Flexible or flex fans reduce the pitch in their blades at high speeds, thereby limiting airflow and lessening engine horse-

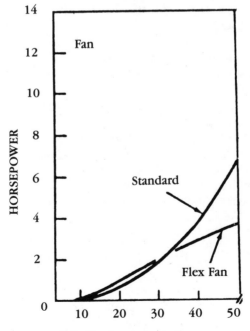

5-2 Flex fan vs. standard.

power requirements. When a vehicle is moving at high speed, it usually gets all the cooling it needs from onrushing air passing through the radiator. By reducing blade pitch at high speed, flex fans allow outside air to do most of the cooling and save gas by easing engine output. A fan clutch can also be used. It unloads the fan at high speeds when it is not needed. Both of these are superior to the fixed-type fan. Mechanical slip-clutch drive and electrical drive fans are also available. These enable the driver to manually control the fan and use it only when necessary. Figure 5-2 shows the superiority of flex fans over standard fans at high engine revolutions/per/minute. Remember, the fan and water pump consume 4 to 5 percent of the total engine-power output.

Plastic wheels weighing 2-3 pounds—compared with standard steel wheels at 21 pounds—could soon be available on new cars. They would have the effect of lowering the total vehicle weight by 100 pounds (this includes the spare, too). Less weight equals less gas used, so if you ever find yourself faced with the option, go with the plastic—all other things being equal, of course.

Fuel injection is usually more mileage-efficient than carburetion for a given vehicle.

In general, a diesel is much more efficient than a conventional engine.

"The smallest engine that provides adequate performance for your needs [acceleration, hill climbing, trailer towing] will also provide you with the best fuel economy. In general, a 10 percent increase in the size of the engine increases fuel usage by 6 percent. Diesel engines can provide as much as 25 percent increase in fuel economy over the same size gasoline engines."

U.S. Department of Transportation

A carburetor degasser is a device used to shut off the fuel supply when a high-manifold vacuum is present, such as during deceleration. Up to 10 gallons of gas per year can be saved by using this device. A few carburetors, such as the Stromberg NA-

75G, have one built in. Since the engine doesn't need gas during deceleration, why waste it? That's the theory behind the degasser. It may become standard equipment on all cars in the very near future. If you're lucky, you might live in an area of the country where one can be bought.

A light-colored vehicle will be slightly more ecomonical in hot climates. Why? Light colors reflect the sun, keeping the car cooler. The air conditioner doesn't have to work as hard to keep the car cool and, conversely, you save gas. Remember this when you purchase your next car.

Would you believe a vinyl top can reduce your highway gas mileage figure? It's true. A vinyl top adds extra resistance to smooth airflow over the top of the car, and it could cost you 1/2 mile per gallon. A sun roof acts in the same way, especially when it's open! Be aware of these popular options that cost you by lowering highway mpg.

A year or so ago I began using an item called the Auto Economizer on my vehicles. It's a patented filter/condensator that fits neatly into the PCV line of any car with a PCV system and has a useful life of about 30,000 to 40,000 miles.

The unit filters out particulate matter including sulfur, carbon, varnish, acids, resins and other high-molecular-weight materials *before* they can be reintroduced into the engine through the PCV line. This filter also contains a water condensing grid that collects water vapor from the crankcase and stores it in a separate compartment. When the engine is restarted, a limited amount of this vapor is fed into the engine.

Crankcase fumes filled with impurities are cleansed prior to being reintroduced into the engine and the engine gets clean, filtered air, not dirty polluted air. According to the manufacturer, the filter helps keep the engine and spark plugs much cleaner, while extending the useful life of the oil, oil filter, spark plugs, and emissions-control components. And because it doesn't have to deal with so much contaminated air, the catalytic converter's life is greatly extended.

In addition to these benefits, fuel economy is enhanced, and noxious emissions are reduced. At present, the Economizer is only available by mail. For information, write:

THE MILEAGE COMPANY
Box 40063
Tucson, AZ 85717

WORTHWHILE ADDITIVES, ENGINE TREATMENTS, AND LUBRICANTS

The following section has been reprinted by permission of the New York Times Syndication Sales Corporation from Bob Sikorsky's "Drive It Forever" column.

I have always been a firm believer in adding some type of solid lubricant supplement every time I change my oil. For many years now, I have added either a molybdenum disulfide (moly), graphite, or polytetrafluoroethylene (PTFE) supplement to my oil when changing it. Perhaps the longevity of some of my vehicles and their trouble-free engine performance can be largely attributed to these solid lubricants providing additional engine wear protection while simultaneously reducing internal engine friction. My old Volvo now has over 350,000 miles without a single engine repair, while my even older Volkswagen bus has nearly 200,000 miles on its little air-cooled engine—again, without any major repairs.

Solid lubricants are just what their names imply. They are non-liquid lubricants that are added to the oil (or transmission fluid or rear axle fluid) and enhance the fluid's lubricating and protective properties. These solid lubricants bond themselves to and fill in the microscopic cavities, the metal surfaces inside the units they are added to (FIG. 5-3). The protective film reduces the friction between sliding and rotating parts, enabling the mechanical unit to give better performance with less effort and wear. Picture someone stretching an impervious layer of Saran Wrap over each engine part—they work something like that.

But to work properly, each of these solid lubricants must be colloidally suspended in the oil. If they are not, the particles of solid lubricant tend to *agglomerate* or separate and settle out to the bottom of the container or, once in your motor oil, will settle out in the crankcase. That's why I have always cau-

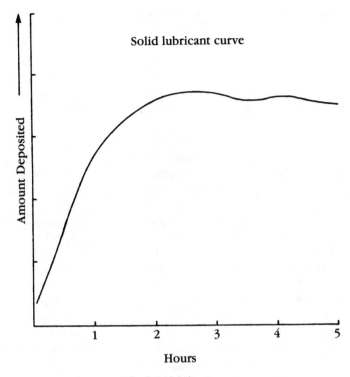

Solid lubricant curve

5-3 Solid lubricant curve.

tioned against using any product that tells you to "shake well" before using. If you have to shake the can to mix the product, what's going to happen to it once it's inside the engine? You guessed it—a lot of it will settle out where it does little good and can do a lot of harm. Unless you are strong enough to shake your whole car before using it, stay away from these products, even if they do contain moly, graphite, or PTFE.

A colloidal suspension, on the other hand, means that the particles of moly, graphite, or PTFE are milled to super-small particle size, usually less than 5 microns, and preferably about 1 micron. (A micron is one millionth of a meter.) These small particles are easily suspended in the carrier oil and will not migrate to the bottom of the container or the bottom of your engine.

Numerous lubrication and *tribology*—or the study of the relationships of friction, lubrication, and wear—studies vividly demonstrate that all three of these products can increase fuel economy and reduce wear of the engine, transmission, and rear axle or transaxle. Unlike some oil "additives," they are chemically inert and won't disturb the carefully blended additive package of the motor oil itself. In fact, they enhance it.

These are especially valuable during winter because they make lubrication-starved cold starts much easier. By placing a slippery coating over and between moving metal parts, the engine cranks or turns over much easier because it has less friction and drag to overcome. And the ravages of cold-start wear, during the exact time when lubrication efforts of the motor oil and the engine's oiling system are marginal, is greatly reduced.

Each of these solid lubricants have very low coefficients of friction. They are much more slippery than motor oil. Thus, when added to the oil, they add their low coefficient of friction to the new mixture. The lower the coefficient of friction, the less friction produced between the parts in question—in this case, the moving mechanical lubricated parts of your car.

So why don't oil companies add these solid lubricants to their oils at the processing plants? Economics. They are expensive. Of the three, moly and graphite are the least expensive; PTFE-type compounds are the most expensive.

The finished, consumer-usable products range in price from about $6 up to $35. Some are added with each oil change, others less frequently. Each can offer its own array of benefits to the car owner looking for economy of operation, extended engine, transmission and axle life, more miles per gallon, and money saved because of reduced operating and repair costs.

I have had personal experience with a number of PTFE, moly, and graphite engine treatments, and as with any type of product, I found that some are better than others. If you would like my list of preferred engine treatments, send a self-addressed envelope to:

THE MILEAGE COMPANY
Box 40063
Tucson, AZ 85717

Tests conducted independently by the Ethyl Corporation, Automotive Research Associates and Loughborough Consultants Limited have shown that a 2-12 percent improvement in gas mileage can be expected if molybdenum disulfide MoS_2) is added to the crankcase oil. Moly oil treatments have been around for years, but the recent gas shortage has brought their remarkable lubricating qualities to the public eye. MoS_2 treatments are now available at many auto parts stores under various commercial names. Molybdenum disulfide, when combined with regular motor oil, dramatically enhances its lubricating qualities and cuts internal engine friction, a major cause of gasoline consumption. Moly will also tangentially reduce fuel consumption due to its effect on increased cranking (starting) speed at low temperatures. The reduction in friction is beneficial during the entire warm-up period, when fuel demands are higher.

A U.S. Department of Transportation report shows that a driver who averages 12,000 miles a year at 12 mpg, with gas cost set at $1.50 cents a gallon, will save $69 a year on gasoline alone by using MoS_2 in his oil. This figure would be much higher with a more realistic mileage-traveled figure. Additional savings are gained through extended oil life, lessened oil consumption, and reduced engine wear and tear. A can of MoS_2 with each oil change will pay for itself many times over. It's a treatment you can use right now to start enjoying the extra miles per gallon.

Everyone is familiar with the type of oil additive that comes out of the can looking (and pouring) like pure honey. So many people use this type of oil additive and swear by it that to knock it would be a national crime. So I won't. Can this type of oil additive improve performance and gas mileage? If your engine is worn a bit and using some oil, then these viscosity-extending additives can help increase engine compression, resulting in better performance and improved mileage. But according to a recent consumer study, a driver can get the same effect by using heavier oil (40W-50W) oil in the first place.

There are specially formulated moly, PTFE, graphite, and synthetic-fortified greases made for packing front wheel bearings.

They offer a considerable friction-reduction advantage over standard wheel bearing greases and that means additional fuel savings.

Besides specially formulated bearing greases, manual and automatic transmission and rear axle treatments containing moly, PTFE, graphite, or synthetic material are also available. Adding these products to their respective mechanical units reduces friction and operating costs, cuts the wear of these expensive parts, and boosts fuel economy.

If you live in a section of the country that gets severe winters, see if you can find a service station that offers a lighter-grade differential and gear lubricant. Under constant freezing conditions, many gears never get a chance to work freely because the lubricant never becomes fluid enough. A lighter grade will warm faster, enhance gear action, and allow the car to move more easily.

If your engine is beginning to use oil and show signs of wear and reduced fuel economy, you might consider one of the metal-plating, tablet-form treatments that are commercially available. Selling for about $10-$15 per treatment, these tablets are placed in the gas tank where they eventually dissolve. The suspended colloidal particles are then carried to the engine, where they plate worn areas. If your engine is not too far gone, these tablets could be of some help in restoring lost power and economy.

Weight of engine oil has been shown to affect gas mileage. A lighter or lower-numbered oil will flow easier and lubricate better when cold than will a heavier, more viscous oil. It also takes less engine power to pump a thin oil than a thick one. Therefore, a 10-weight oil will give better gas mileage than a 40-weight oil. By switching to a lighter oil, especially for winter driving, you allow your engine a better chance to produce top mileage.

If you are of the old school ("heavy oil protects better") and feel more comfortable with a heavier oil, then multigrade oil is the answer. A 5W-30 or 10W-30 oil is relatively thin when it is cold, and then thickens as the engine warms and demands extra protection. You gain economy by having the thinner, easy-flowing oil for cold-engine temperatures when friction and gas consumption are their highest. Whatever we do to improve cold-engine economy pays off in double savings because a cold engine uses twice the amount of gas doing the same work as a warm one. A quality multi-viscosity oil is one way to help cut the cost of cold-weather and cold-engine operation.

Do you know what the words "Energy Conserving" or "Energy Conserving II" stand for on a container of motor oil? According to the American Petroleum Institute (API), "Engine oils categorized as Energy Conserving are formulated to improve the fuel economy of passenger cars, vans, and light-duty trucks. These oils have produced a fuel economy improvement of 1.5 percent or greater over a standard reference oil in a standard test procedure. Oils meeting this requirement display the Energy Conserving label in the lower portion of the donut-shaped API Service Symbol."

Oils that carry the Energy Conserving II label offer even greater fuel economy rewards. They have been tested, and show a fuel economy improvement of 2.7 percent or greater when tested the same way as the Energy Conserving oils. In other words, when compared to a standard reference oil without an Energy Conserving designation, they will improve gas mileage by 2.7 percent or better.

In its owner's manuals, Ford Motor Company says, "For maximum fuel economy benefits, use an oil with the Roman Numeral 'II' next to the words 'Energy Conserving' in the API Service Symbol." That's not a bad idea, no matter what kind of car you drive.

In my opinion, each car owner in the United States should be using an Energy Conserving II motor oil. If you are now getting 20 mpg, switching to an Energy Conserving II motor oil will up your mileage to 20.54 mpg. Although that doesn't sound like much at first, consider that if you have a 15-gallon tank, it means you will get an extra 8.1 miles free on each tankful simply by using the EC II oil. That's like getting a free half gallon of gasoline or more with every fill-up. Because EC II oils cut friction, another bonus you'll receive is reduced engine wear.

The lower portion of the donut-shaped API Service Symbol that appears on all oil containers is reserved for either the "Energy Conserving" or "Energy Conserving II" designations. If an oil has no energy-conserving properties, that area will be blank (FIG. 5-4).

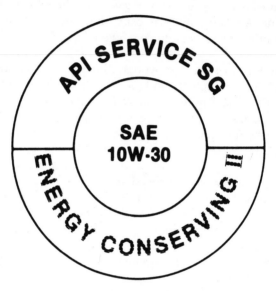

5-4   An example of the API Engine Oil Service classification symbol.

What about liquid gasoline additives? Can they help improve gas mileage? There are many of these on the market, and their main value as a mileage aid may be that they help clean the carburetor and fuel system and lubricate the top cylinder area, resulting in slightly increased engine efficiency.

Although most gasoline additives are a waste of money, owners of fuel-injected vehicles might be wise to occasionally add a container of fuel-injector cleaner to the gas tank. The cleaner is extra assurance that the injectors will be deposit-free and operating properly. A can every third or fourth fill-up is a good preventive maintenance practice.

Can synthetic oils save you gas? By reducing internal engine friction a substantial degree over conventional oil, synthetic oil promotes better mileage in the majority of cars. The mileage improvement varies greatly and is directly related to the condition of each individual engine. Realistically, a driver can expect anywhere from 10-20 extra miles for each 20-gallon fill-up. The high cost per quart (around $4) is the only negative factor. Better performance, less frequent oil changes, better mileage, and prolonged engine life are the plus factors. For your purposes, synthetic oil can and does improve gas mileage. Most synthetic oils are also rated as Energy Conserving.

Carry a siphon in the emergency kit in your car. Although it won't do a thing to improve your gas mileage, it might be just what the doctor ordered to get your car moving again if you run out of gas or are running on empty. That's providing, of course, you can find a sympathetic motorist to "lend" you a gallon or two.

In a booklet about saving gasoline I wrote for the State of Arizona called "Ease The Squeeze," there was a pressure-sensitive sticker with the words "THINK ECONOMY" printed on it. Readers were advised to place this sticker on their dash to remind them to think and drive economically.

If you would like to receive up to five free pressure-sensitive stickers, send an SASE to:

THE MILEAGE COMPANY
Box 40063
Tucson, AZ 85717

Good maps or a road atlas should be every economy driver's companions. They are especially valuable when traveling in a strange town or state. Check the maps before you start out on a trip, whether it will be across town or across the country. A glance at a map might reveal a better or shorter route to your destination. With a map in hand, it's easier to plan your route for the most efficient use of your automobile. A good set of maps will pay for themselves many times over in fuel savings and less wear and tear on your car.

An item that more pickup truck owners are using to save gas is the open mesh rear tailgate. These net-like "gates" take the place of the tailgate on a pickup. The normal tailgate on a pickup truck acts like a brake and holds the truck back at highway speeds. Wind rushing down and over the cab of the truck and into the pickup bed runs smack into the tailgate. This extra wind resistance the truck must overcome is paid for in extra gasoline used.

The net-like rear gate allows the wind to flow through the spaces in the net and, thus, greatly reduces the drag on the vehicle. Such a gate could help if you do a lot of highway driving. Of course, you could always lower the tailgate and get the same results, but the pickup bed must be empty if you do.

Chapter **6**

Keys to
maximum mileage

The five keys to maximum mileage are to: inspect, clean, adjust, alter, and repair. Whether you do it yourself or take your car to a service station, practicing these methods can increase your gas mileage.

THE CARBURETOR

On pre-computer vehicles, the fast-idle cam adjustment, which controls cold-engine idle speed, should be set on the lowest cam step that allows sufficient cold engine rpm to prevent stalling. Too high cold-idle speed rapidly eats gas. Keep it as low as possible. In summer, the fast-idle cam screw can usually be set at the lowest cam step for additional gas savings.

Because it might be necessary to increase idle speed to compensate for heavy air conditioner use, remember to turn the idle down when the air conditioner season is over. Without the extra pull of the compressor, engine idle speed can be cut by at least 100 rpm.

Some drivers will disconnect and totally bypass the carburetor accelerator pump in their quest for better mileage. Though this will eliminate the gas-wasting squirts of the pump, it leaves the driver with little recourse when fast acceleration is necessary. You can save some gas by doing it, but I don't recommend disconnecting the accelerator pump, because if you have to accelerate in a tight situation, you won't be able to do it fast enough.

You can get better mileage by shortening the carburetor accelerator pump stroke. Move the accelerator pump rod end to a different hole so that it shortens the pump stroke. This lessens the amount of gas pumped into the carburetor each time the accelerator is depressed. Expect a slight reduction in top speed with this easy adjustment.

A decrease of $1\frac{1}{4}$ percent in the air/fuel ratio occurs for every 1,000-foot rise in elevation. So for best economy, carburetor richness should be reset if you move from a lower to a higher altitude, or vice versa. At higher altitudes, where air is thinner and less available, the carburetor must be set leaner to compensate for the less dense air. A slightly richer setting is desirable at lower elevations, where air is more dense.

If you smell raw gas when starting a cold engine, it usually means that the automatic choke is set too rich. Lean the choke setting to the point of easiest starting. This also is usually the most economical setting. Some newer cars have electrical assists on the choke to help open and close the valve. Check to make certain the choke is functioning properly and the choke valve is fully opened when the engine is warm.

The choke adjustment is extremely important. A too-rich automatic choke wastes copious amounts of fuel. In some cases up to 95 percent (you read that right) of the fuel supplied to the engine is unburned during the interval when the choke remains closed (or set too rich). That means that a car capable of 20 mpg when warm may only get 1 or 2 mpg when running with a too-rich choke.

On carbureted cars, a sticking or maladjusted automatic choke is the number-one cause of poor wintertime cold-engine performance and fuel economy. Don't leave home without a properly working choke.

Many carburetors are equipped with a *hot-idle compensator*, a thermostatically controlled valve located in the upper throat of the carburetor that is usually visible when the air cleaner is removed. During long periods of idling with a hot engine (for instance, in summer rush-hour traffic), the fuel in the carburetor bowl can become hot enough to vaporize. These vapors can enter the carburetor bores, mix with the idle air, and be drawn into the engine—causing an extremely rich mixture that can cause the engine to stall. The hot-idle compensator opens under these conditions, permitting additional air to enter the manifold and mix with the rich fuel vapors, providing a more combustible mixture. Extremely rough idle operation and engine stalling are avoided. If your car has a hot-idle compensator and you have any doubts concerning operation of the valve, replace it.

The carburetor float should be checked for proper alignment. It should move freely up and down and not scrape or hang up on the sides of the bowl. This will allow excess fuel to enter the bowl. Remove the float and shake it to make sure it isn't filling with gasoline. A gas-filled float will sink, opening the needle valve and allowing continuous wasteful passage of gasoline.

Check the carburetor needle valve and seat for wear and for dirt particles that may be trapped there. A needle valve that doesn't seat properly allows extra gas to leak into the bowl and create an overrich mixture, often culminating in severe flooding and wasted gas. Replace the needle valve if it shows any signs of scarring or wear.

On four-barrel carburetors, disconnect the secondary throttle linkage. If you're economy-minded, you won't need the extra power provided by multiple throttling.

A sticking carburetor accelerator pump is an all too common gas thief. If your car hesitates or the engine almost dies when the accelerator is pushed down, you can bet the accelerator pump is at fault. This becomes more apparent when the gas pedal is suddenly floored—the car hesitates, coughs, and then gradually picks up speed. Accelerator pumps are not expensive and are relatively easy to install. They can be purchased singly or in carburetor-rebuilding kits.

Automatic choke settings do not have to be as rich for summer as for winter use. Lean the choke when warmer weather arrives, using the leanest possible setting that permits easy cold-engine starting. When the engine has warmed, check the choke valve to see that it is fully opened; if it isn't the setting is still too rich.

Lowering the carburetor float level approximately $1/16$ inch below factory specifications is another way to enjoy a slight increase in gas mileage. Don't lower it more than this amount, or you can starve the carburetor.

If your car is a rough idler, don't increase the idle speed to overcome the roughness—that costs you gas. Some of the first cars built with antipollution equipment are notoriously poor idlers, and, unfortunately, there isn't much that can be done about it. Set the idle at the lowest rpm where the engine won't constantly stall.

If you don't think idle speed has much of an effect on gas mileage, just look at FIG. 6-1. Reducing idle speed from 900 rpm to 400 rpm increases deceleration gas mileage by a whop-

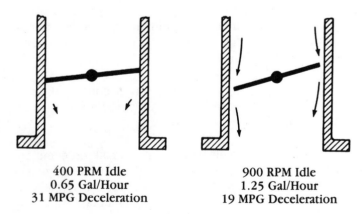

400 PRM Idle	900 RPM Idle
0.65 Gal/Hour	1.25 Gal/Hour
31 MPG Deceleration	19 MPG Deceleration

6-1 Idle speed and its effect on gas mileage.

ping 12 miles per gallon and cuts in half the amount of gas needed to idle the engine for one hour.

On newer cars, idle speed is controlled by a computer which receives its input from an electronic idle speed sensor. On these vehicles there is little if anything the owner can do to adjust the idle speed, nor should any attempt be made to try to "outguess" the computer.

Many cars have a tendency to *diesel*; that is, the engine continues to run after the ignition is turned off. This not only costs you gas dollars but is harmful to the engine as well. Some cars are equipped with an idle-stop solenoid, usually located near the carburetor, that prevents this condition. The solenoid holds the throttle linkage at idle while the ignition is on; when the ignition is turned off, it closes the throttle to prevent further idling. If your car has a dieseling problem, check to see if it has an idle-stop solenoid, and have it replaced if it's not working right.

When replacing carburetor gaskets, make sure that the new ones have the exact same cutouts and holes as the old ones. A gasket that is not exactly correct may plug a vital vacuum or vent passage and adversely affect performance.

Check the idle-air bleed hole on top of the carburetor to be certain it is free of obstructions. A clogged idle vent will hurt mileage by enriching the idle air/fuel mixture.

You can conscientiously follow every suggestion in this book, but if the carburetor air/fuel ratio is too rich, your efforts won't be totally rewarded. The air/fuel ratio must be set on the lean side to obtain best mileage. It may be the single most important adjustment you can make to improve gas mileage and it is critical to have it checked and any necessary changes made. One way to do this is to take your car to a garage that has an exhaust gas analyzer. This machine will "read" your car's exhaust, and tell you if the air/fuel ratio is correct or needs changing. If the ratio is too rich, the garage will probably suggest changing either jets or metering rods to bring it into a more economical range. Some overrich carburetors, when leaned out, will improve economy up to 5 mpg. Make sure you don't overlean (an air/fuel ratio of 14.7:1 is usually excellent) because you can damage valves and spark plugs and get erratic engine performance if the carburetor is starved.

When you lean out the carburetor, make certain the entire ignition system is in good shape because it requires a hotter spark to ignite a leaner mixture (FIG. 6-2). Figure 6-3 shows how air-fuel ratios affect efficiency and emissions.

FUEL-INJECTION SYSTEMS

Most newer cars today are equipped with fuel-injection systems. When the original edition of this book was written, fuel injection was found exclusively on slick, imported sports sedans.

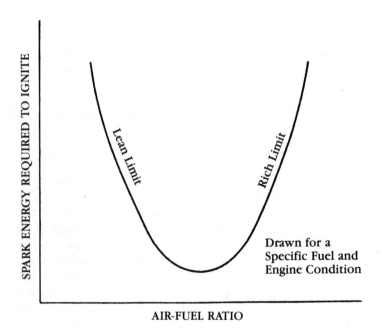

6-2 Energy required for ignition.

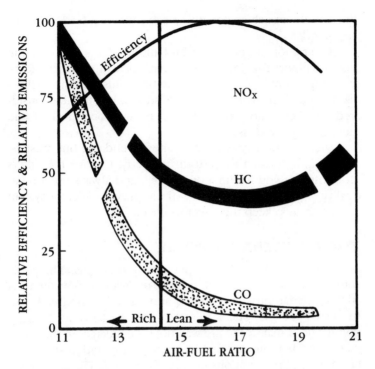

6-3 Effect of air-fuel ratio on efficiency and emissions.

Fuel injection is rapidly replacing the faithful carburetor as the main fuel-supply device. It is more efficient and economical, and it allows the engine to perform closer to its designed potential.

Most fuel-injection systems electronically inject fuel into the engine in a variety of ways. Current systems are called *single point, throttle body,* and *multiport* or *multipoint* injection.

Single point means there is just one point where the fuel is injected. *Throttle body* means that the point of injection is on the throttle body, a unit that sits on top of the intake manifold. In both systems, fuel is injected either through one or two injectors into the throttle body where it is mixed with air and fed into the intake manifold which delivers it to the engine's cylinders.

Multipoint or multiport means there is more than one point where fuel is injected. Typically, an injector is located directly above each engine-intake valve where it feeds fuel into the cylinder.

A variation of multiport and multipoint systems is the sequential-port fuel-injection system. Instead of all cylinders receiving a shot of fuel at the same time, this system synchronizes the delivery with each intake stroke. It is usually considered the best, most precise, and most economical of the fuel-injection systems.

Fuel-injection systems are quite dependable but must be kept sparkling clean to perform properly. Change fuel filters more often if you live in a dusty area, and always use a high-quality, high-detergent fuel with a proven injector-system cleaner to help keep the injectors clean.

THE ACCELERATOR PEDAL

A sticky, balking accelerator pedal will waste gas. The pedal should have a smooth but firm up-and-down movement. A sticking pedal must be pumped and jiggled to be freed, and this costs you gas. Check to see that the pedal is not hanging up on a floor mat or the fire wall hole. All linkages in the accelerator pedal/carburetor/fuel injection configuration should be clean, free of obstructions, and easily moved. The accelerator pedal is your connection to the engine and always responds in exactly the way the foot commands it. It must work easily for the driver to have good feel.

A too-easy-to-push accelerator pedal also wastes gas. It should offer moderate resistance. If it pushes too easily, install a stronger or shorter return spring. There are units on the market selling for up to $20 that automatically resist foot pressure, forcing you to use less pedal and save gas. A shorter or stronger return spring, selling for about $1.00, will do basically the same thing.

On the other hand, an accelerator pedal that has too much natural resistance will waste gas, too. It is difficult to get good feel with too much return-spring tension, and the car will tend to lurch ahead whenever foot pressure is applied. In this case, lengthening the return spring will cure hard pedal and eliminate jerky acceleration.

Some economy-minded drivers place a wooden block under the accelerator to prevent it from being floored. Although the motivation behind this practice is accurate enough, it can be dangerous. In an emergency, full power may be needed but won't be there because the wood block stops the accelerator. If you now practice this method of saving gas—discontinue it, please. With practice, the egg or apple tricks previously mentioned will give the same results with no danger involved.

FUEL PUMP, FILTER, GAS LINES, GAS TANK

A properly operating fuel pump is absolutely necessary for top mileage. Excessive fuel-pump pressure will force more gas to the carburetor than it needs and can bend or damage the float arm. Too little fuel-pump pressure, on the other hand, can starve the carburetor, cause missing, overburden the engine, and eventually result in major damage. If you have any doubts about your fuel pump, have it checked for proper pressure (usually around three to five pounds) at a reliable garage. Heavy, raw gas odor may indicate a ruptured fuel pump diaphragm.

Make certain that your car has a fuel filter; the in-line paper element kind is best. By filtering out foreign particles before they reach the carburetor, the fuel filter ensures clean, steady gasoline delivery. On older cars, change filters ever 5,000-10,000 miles or when they are visibly dirty.

Vehicles with fuel-injection systems are particularly susceptible to dirt in the gasoline. While a carbureted vehicle can be a bit forgiving of dirty fuel or can operate a bit longer with a dirty fuel filter, not so their fuel-injected cousins. A fuel-injected vehicle must have a clean fuel filter to operate efficiently, so be certain you change yours at least as often as the manufacturer suggests, and more often if you commonly drive in dusty or dirty areas. Poor fuel economy in a fuel-injected vehicle can often be traced to a clogged or dirty fuel filter.

Be certain your vehicle has the proper type of gas cap. Most older cars use the vented variety, while newer cars with evaporative emissions controls use a non-vented cap. The wrong cap on a car can cause poor performance, increase emissions, and reduce fuel economy.

An unvented cap on a vented system could even cause the gas tank to cave in. A vented cap on an unvented system creates unnecessary pollution and affects the engine's performance. Gas caps are available at any auto parts store or service station.

Make sure that your gas cap fits snugly. Any rain or splashed water that might seep into the tank can cause misfire and engine-power loss and create a lot of unnecessary pollution.

There should be no leaks in the fuel system. Check the gas tank, fuel lines, fuel pump, fuel filter, and carburetor for signs of escaping gasoline. Points where metal gas lines are connected to rubber ones are especially susceptible. The juncture of the gas-filler neck with the tank is also a likely area. All carburetor screws and nuts should be snug and gaskets in good order. A bad gasket on the carburetor accelerator pump will allow gas to escape through the vent hole and will wet the outside of the carburetor. If you smell gas and suspect a leak, park the car on a clean, dry surface and check for wet spots underneath the car for a clue to the area the leak is in. Sometimes a leak will occur only if the engine is running with normal fuel-pump pressure. Therefore, be sure to check with the engine running also.

SPARK PLUGS AND WIRES

Spark plugs, one of the most important items that affect gas mileage, should be cleaned and gapped on older cars at least every 5,000 miles to ensure the best possible mileage. If any plugs are bad, you should replace them immediately and not try to nurse them for a few thousand more miles. In a four- or six-cylinder engine, just one malfunctioning plug can cut 10 percent off the top of your mileage (FIG. 6-4).

A popular trick used by Mileage Marathon drivers is to widen spark-plug gap an extra .010 over specifications on older vehicles. You will experience some loss of top speed with this maneuver, but it is worth it if you are economy conscious. Some of the newer cars with electronic ignitions have plugs

- One plug in a V-8 engine misfiring half the time at 55 mph reduces gas mileage by 7%.
- Two plugs misfiring reduce gas mileage by 20%.

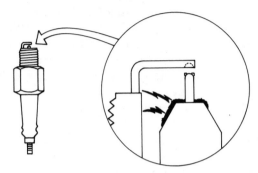

6-4 Effects of spark plug misfiring.

that gap up to .075. On these, don't widen the gap further but keep it at the original specs.

When cleaning and gapping plugs, make sure that the electrodes are filed to a nice, sharp, square edge. Spark has a tendency to cling to sharp edges in contrast to rounded ones. Crisp, squared-off electrodes will permit more spark, better combustion, and better mileage. This is one of the main reasons new plugs improve performance.

A rounded feeler gauge is usually more accurate for setting spark plug gaps than the flat type.

A slightly loose spark plug will rob the engine of full compression and noticeably affect gas mileage. Seat all plugs properly and torque them to specifications. If the plugs require O ring seals, they should give a little to ensure a proper seal between the plug and engine block. If they are flat, compression leaks may occur.

Newer cars that use computer controls, high-energy ignitions, and unleaded fuel can go much longer between spark plug changes and adjustments. In fact, spark plugs on newer cars can last 30,000 miles without negatively affecting fuel economy. On these cars, spark plugs rarely need cleaning or gapping. However, the plugs should be inspected at each tune-up and changed, gapped, or cleaned as required.

Copper- or platinum-tip plugs last longer, stay cleaner, and give better service and better spark than conventional spark plugs. They are well worth the investment of the few extra dollars they cost. In the long run, they will permit the car to run more efficiently for longer periods of time, and they require much less attention than conventional plugs. Their use is highly recommended in both newer and older vehicles.

Spark plug wires should be inspected periodically for signs of cracking, burning, wear, and oil or grease contamination. Ignition wires are the arteries of your engine, and like those in your body, they must be in good condition for unimpeded electrical flow.

Keep spark plug wires separated. Grooved plastic spark plug looms or separators are usually provided for this purpose. If your car doesn't have spark plug looms, it should. They help prevent cross-firing or shorting of the spark from wire to wire, as well as the consequent engine miss that could occur if the wires are too close to each other.

Do not tape or tie plug wires together, as this can cause inductive cross-firing across the wires and upset the firing order of the engine.

Plug wires that are too long will build up extra resistance and cut the intensity of the voltage delivered to the plugs, reducing their ability to detonate the fuel charge. If your plug wires are sloppy and longer than necessary, cut them to the shortest practical length that will extend between plugs and the distributor cap. The less distance voltage has to travel, the better its quality will be; thus, short plug wires improve engine economy. Make certain, also, that the high-tension wire between the coil and the distributor is as short as possible.

Sometimes an old or malfunctioning ignition switch can be

responsible for hard starting and drop in engine efficiency by impeding electrical flow in the primary ignition circuit.

Spark plugs must be in the correct heat range to get top mileage. If the heat range is too low (cold plug), the plug will foul, cause rough-engine idle, and allow carbon deposits to build up. If the heat range is too high (hot plug), it can burn valves, crack plug insulators, and destroy electrodes. As a general rule, it is best to stay with the type of plug recommended by the car manufacturer. However, additional gas savings can be realized by installing a slightly hotter plug if you do a lot of city driving, or a range colder plug if you do mostly highway driving. Let plug condition dictate if you should change, but don't go up or down more than one heat range or plug number at a time. If plugs are heavily carbon- and oil-fouled, that's usually a sign that they are too cold and won't allow complete combustion. Metal deposits on the electrodes, cracked insulators, whitish deposits, or eroded electrodes signal that the plug is too hot and a colder range is desirable.

THE DISTRIBUTOR

Distributor breaker points that are worn, pitted, dirty, oily, misaligned, or improperly gapped are a major cause of bad gas mileage. Point gap or *dwell* should be checked at least every 5,000 miles because the distributor shaft rubbing block that causes the points to open and close will wear, narrowing the gap. A new set of points is relatively inexpensive, so if in doubt, replace them.

Do you seem to be constantly replacing the ignition points on your older car? Points play an important role in the fuel economy equation and must be in good condition and set to the proper gap. Many drivers—and mechanics, too—overlook an all-too-common cause of point degradation: a leaking distributor shaft seal. If this seal is not in good condition, fumes from the crankcase will find their way inside the distributor cap. The combination of oil and ionized air in the area of the points can cause a rapid degradation in point life, and fuel economy will suffer as a result.

McDonnell Douglas's fleet-maintenance department had a policy of removing the distributor from the engines of their vehicles and replacing them with overhauled units to ensure an accurate fit of seals, bearings, and gears. According to McDonnell Douglas, this policy has contributed significantly to an increase of fleet fuel economies.

You don't have to replace your distributor, but you can replace the shaft seal if there is any indication of oil residue inside the distributor cap.

Some distributors are equipped with an electrical solenoid that allows additional spark advance during starting. If this is not working properly it will make starting hard, if not impossible. Check all connections to and from the solenoid for tightness, and replace the solenoid if there is any doubt about its performance.

On other vehicles, distributor vacuum advance may be controlled by means of a thermal vacuum switch (TVS) that senses engine water temperature. When the temperature rises, spark is automatically advanced to smooth engine performance. Check all hoses leading to and from the TVS for leaks.

Some cars have a switch located on the transmission that controls spark advance as a function of gearing. When the vehicle is in high gear, the spark advances; when in first or second, it retards. If this switch isn't functioning in the proper gear ranges, it will cut into gas savings. If in doubt, have it checked.

Be certain that the distributor rotor isn't worn, cracked, or dirty, and that it fits snugly on the distributor shaft. The rotor is the point of spark transference and must be in good condition if a hot spark is to make it to the cylinders.

Check the outside and inside of the distributor cap to make sure it is not cracked, burned, dirty, or wet. Look closely at the electrodes inside the cap for signs of excessive wear. Just a hairline crack, a spot of grease, or a bit of condensation inside the cap can make the spark diffuse over the entire cap surface, causing backfire and engine miss.

A weak or leaking ignition condenser (located under the distributor cap) is an often-overlooked thief of good mileage. A malfunctioning condenser won't allow the coil to deliver peak voltage to the spark plugs. Bluing or excessive pitting of the distributor contact points usually indicates a faulty condenser.

The distributor automatic advance must work freely. Its job is to advance spark to the cylinders during acceleration and high-speed operation. If a spring or diaphragm in the automatic advance mechanism is broken, engine miss and a noticeable drop in economy will result. The distributor shaft should turn with slight hand pressure and then return to its original position when the pressure is released. The automatic advance must function properly to ensure maximum mileage at lower speeds.

One of the most popular ways to increase gas mileage is to advance the ignition timing 3-5 degrees over factory specifications. Factory-set timing is, at best, a compromise between the different types of driving encountered throughout the country. Advance it a few degrees and enjoy extra mileage from your car.

Although changing the timing can be done rather easily, it is best to let a properly equipped garage handle it. Timing that is advanced too far will do more harm than good, so don't chance it unless the proper equipment is available. Power tuning, or "timing by ear," is out. Through the years it has probably created quite a few problems, and the driver who tries it usually ends up at a garage having the timing reset. Advanced timing, not more than 5 degrees, is a legitimate, effective way of boosting mileage. Economy Run drivers have done it for years. If not overdone, it can boost gas mileage by 1-2 mpg, depending on engine condition, driving habits, and terrain. However, I suggest that this procedure be used only on non-computer-controlled vehicles.

For the do-it-yourself tune-up advocate, one of the most important items you can recognize is the one-way timing/dwell relationship. When you change the setting of the distributor points (the *dwell angle*), you automatically change the ignition timing; but if you change the timing, the dwell angle doesn't change. For this reason, it is important that the dwell angle be set first; then the timing can be adjusted. This procedure ensures the correct timing/dwell relationship and proper engine operation.

THE COIL

Correct coil polarity is essential if you want top economy. This often-overlooked area can be a cause of very rough engine performance and mileage loss. A coil with reversed (or positive) polarity requires up to 50 percent more voltage to fire a spark plug and places a tremendous burden on the ignition system. If you suspect reversed coil polarity, have it checked at a garage or do it yourself by removing an ignition wire from one of the spark plugs while the engine is idling. Hold the end of the plug wire close to the top of the plug while inserting the end of a lead-tipped wooden pencil between the end of the wire and the plug. If the spark fires on the plug side of the pencil, polarity is correct; if it fires toward the ignition wire, polarity is reversed. To correct polarity, simply reverse the primary wires at the coil.

THE BATTERY

A fully charged battery is a must if you are trying for maximum gas mileage. No Economy Run driver would be caught without one. A battery in top condition delivers a hotter spark to the entire ignition system, allowing better combustion. A charged battery also signals to the voltage regulator, which in turn keeps the generator or alternator output at minimum, preventing horsepower loss. Clean terminals, sound cables, and cells filled to the proper level all help maintain a battery at peak efficiency.

Corroded or loose battery terminals can make the alternator work overtime trying to keep the battery charged. Corroded terminals = more resistance to electric flow = poorer electrical current = more alternator work = less mpg.

NOTE: All connections in the primary and secondary ignition circuits must be clean and tight to ensure maximum voltage and proper operation of each individual unit in the circuit.

THE ENGINE

Although I'm a believer in a clean engine, it isn't advisable to power-spray wash the newer engines to keep them clean. Water where it isn't supposed to be can give a modern engine fits. Newer cars are laden with electronic sensors that control operation of the engine and, even though most electronic sensors are fairly waterproof, H_2O has a way of infiltrating waterproof places—especially when it has a lot of pressure behind it and a lot of detergent suds in it. Mess up even one sensor, and you could pay dearly in fuel economy. Keep engine surfaces clean by wiping them occasionally with a cloth moistened with a mild cleaning solution.

Worn piston rings will cause low engine compression and excessive oil and gas consumption. New rings require a major overhaul, which these days is very expensive. If you plan on keeping your car for a few more years, the cost of the overhaul will at least be partially taken up in the extra gas savings you will gain from the new engine.

Cylinder head bolts should be checked for tightness to prevent any chance of compression leaks around the head gasket.

Don't forget to check the draft tube located underneath the engine in many older cars. It can become clogged with engine and road dirt, and, when restricted, it hinders the engine from breathing normally and functioning at top capacity.

A bad water pump not only won't provide adequate coolant to vital engine areas but will also make the engine work harder to turn it because of bad bearings and scraping rotor blades. To improve water-pump life, turn on the air conditioner only while the engine is idling. This places less initial strain on the water pump bearings as the air-conditioner drive belt is engaged.

Some car owners will remove the air conditioner drive belt during the winter months and go a bit further on each gallon of gas. The only disadvantage here is that the unit should be run occasionally to help keep the seals pliable, and this involves replacing the belt a time or two.

In very cold winter climates, you could try running without a fan and enjoy more gas savings. However, keep a prudent eye on the temperature gauge, and if the car is running too hot, replace the fan belt. If you try this, keep the fan belt in the trunk in case it is needed.

By removing the fan belt, you bypass not only the fan but also the water pump. Combined, these two units use up to 8 horesepower when engaged, so it is obvious that considerable gas can be saved by disconnecting them—if weather allows. Figure 6-5 shows water pump/fan power requirements.

All engine-drive belts should be adjusted to proper tension. Air-conditioner, power-steering, power-brake, air-pump, super-charger, and fan belts should be checked periodically. Belts that are too tight will harm bearings and exert a negative influence on gas mileage because the engine will have to work harder to overcome the extra belt and bearing friction.

Burned or sticky valves lower engine compression and can cause extreme power loss. A sticky valve can sometimes be freed by adding a can of top cylinder oil to the gas or to the crankcase oil. If this doesn't work, the valve is probably burnt and must be replaced.

If you have an older vehicle with "soft" valve seats and decide to have an engine overhaul or valve job, be certain to ask the rebuilder to use hardened valve seats in the rebuild. This will allow you to safely use unleaded gasoline in the vehicle without worrying about valve-seat recession.

Mechanical valve lifters should be adjusted every 10,000 miles or as specified by the manufacturer. When adjusting valve-lifter clearances, it is always better to have them a bit loose than tight. Always err on the side of the larger clearance for maximum fuel economy. Lifter adjustment is especially important with some of the small, 4-cylinder engines. A gradual loss of power and economy will occur if valves are not adjusted at regular intervals.

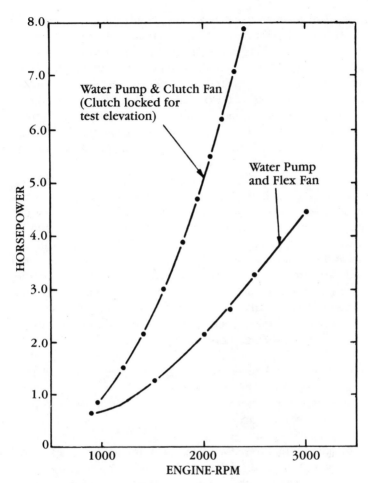

6-5 Power requirements—Water pump and fan.

Although hydraulic valve lifters do not require adjustment, you can extend their life by taking it easy the first minute or so after starting the car. This gives the oil a chance to reach the lifters while the engine is still under light load.

Intake manifold leaks caused by any of the following can upset the air/fuel ratio and result in rough idling and poor economy: loose manifold connections or leaks occurring in intake manifold vacuum lines or at the carburetor flange; loose manifold nuts; distortion or misalignment of gasket surfaces at the intake manifold and carburetor-attaching flange; damaged or improperly installed gaskets; a leak at the juncture of the carburetor and the throttle rod where the rod may have worked loose the vacuum seal. To demonstrate the importance of a sealed intake manifold (no leaks), McDonnell Douglas Corp. was able to improve the fuel mileage of some new automobiles with these problems by as much as 4 mpg simply by resurfacing and properly mating the carburetor to the intake manifold. As you can see, air leaks into the intake manifold can be costly.

A large leak, such as a vacuum hose that has become disconnected, can make itself known by a loud hissing sound or can be detected by visual inspection. Less obvious leaks can be located by squirting a little gasoline or household oil in and around suspected areas while the engine is idling. If there is a leak, the gas will be drawn through it and into the engine, where it will cause an increase in engine speed as it is burned along with the regular fuel. Don't forget to check hoses to vacuum-operated units, such as the windshield-wiper motor, windshield washer, and distributor advance.

Rubber vacuum lines won't last forever. Cracks or brittleness will develop over time. These hoses should be replaced whenever they appear to be aging. Don't wait until a hose starts leaking before you replace it.

It's easy to forget to reconnect a vacuum line after working on an engine. Note all lines that you disconnect by keeping a written account and, with checklist in hand, go back and reconnect the lines when your work is finished. As we have seen, a disconnected vacuum line will play havoc with engine performance and fuel economy.

On newer cars, vacuum leaks will skew sensor readings and provide the engine's computer with "false" readings, which will result in poor economy.

THE TRANSMISSION

Automatic transmissions should be adjusted to shift at factory-recommended intervals for most economical operation. Too much time in low range foolishly wastes gas, and too short a time will cause the car to labor once it has shifted to high gear. Check the owner's manual or ask a qualified serviceperson at what point your car should shift. If it is shifting too early or too late, have it adjusted. The cost is minimal, and gas savings will soon pay for it.

A slipping automatic transmission will eat gas because much of the power sent to the rear wheels is lost. The transmission should shift smoothly and surely. If it hesitates, jerks, whines, or slips its way into a higher gear, have it checked. Gas that should be propelling the car is instead being wasted.

By the same token, a slipping clutch in a standard-transmission vehicle will also use extra gas. A slipping clutch is never fully engaged, and power is lost throughout the entire drive train. If the clutch plate isn't worn too badly, a minor adjustment will cure this. Three-quarters of an inch of free play in the clutch pedal is usually the best adjustment.

THE AIR CLEANER AND FILTER

A clean air filter is another must for peak gas mileage. When the air filter becomes clogged and dirty, the volume of air that is capable of passing through it is reduced, starving the carburetor of air and forcing it to use more gas. Replace the air filter when it is dirty, regardless of how long it has been in use. Remember: In areas where there is a high percentage of dirt roads and blowing dust is common, the filter element will tend to clog up fast and should be replaced more frequently.

Don't try to clean a paper-element air filter. You can do more damage than good with this procedure. Dirt and dust can be removed by blowing the filter out with an air hose or pounding it on the ground, but when the filter is replaced on the engine, the vacuum will suck other particles (now loosened) into the engine and set the stage for hastened wear. Always replace the filter when in doubt. Air filters aren't expensive, and they are cheap insurance against engine wear. A new one will enhance fuel economy.

An easy way to get more efficiency from the air filter element is simply to rotate it 180 degrees. This brings the cleanest part of the filter around to the point where air is drawn through the air cleaner opening (*horn*). By doing this, the unused part of the filter is brought to the point of air contact and is able to filter more effectively and with less restriction, thus increasing the volume of clean air available for combustion.

Drilling a line of holes around the circumference of the air cleaner is another way of supplying extra air to the carburetor. Some drivers will even go so far as to cut the entire side out of the cleaner, leaving only the top and bottom and exposing the entire surface of the air filter element to outside air.

Instead of drilling holes, you can dehorn the air cleaner. This consists of nothing more than cutting off a section of the air cleaner horn or snout so that it has a wider opening, permitting more air to enter.

Look inside the air-cleaner snout to see if it has a diverter valve. This valve monitors the amount of preheated manifold air that can be drawn into the carburetor via the air-cleaner diverter tube. It must be free-moving to ensure correct amounts of outside (cold) and manifold (warm) air. If the diverter valve is stuck in the closed position, the engine will be robbed of the prewarmed air it needs to operate efficiently when it is cold and the extra air it needs when it is warm.

ENGINE AIR DUCTS

The following section has been reprinted by permission of The New York Times Syndication Sales Corporation from Bob Sikorsky's "Drive It Forever" column.

Car been acting up lately? Does it exhibit any of the following symptoms: poor starting, loss of power, rough running, excessive emissions? And you say you had a tune-up not too long ago? Don't know what could be causing it? And to top it off, your gas mileage has dropped too?

Before you give the beast two aspirin, cross your fingers and send it to bed, perhaps you should check one of the most ignored under-the-hood items. In fact, these parts feel so neglected that they go to bed many evenings with tears coming out of their ducts.

Oops! Gave it away. Yep, I'm talking about the carburetor and fuel-injection air ducts. Air ducts come in many sizes and shapes, but they all perform the same function. They route air to your car's engine.

There are two types of air ducts: fresh-air inlet ducts and preheater ducts. Fresh-air ducts supply outside air to the engine when it is warm, and preheater ducts supply preheated air to the engine during the cold-start and cold-engine warm-ups and operation.

Just as all roads once led to Rome, all air ducts lead to the engine air cleaner. The fresh-air duct provides air from some point near the front grill, and preheater ducts get warm air from some point on the intake manifold of the engine.

Both types of ducts look like overgrown hoses that have a corrugated, crinkly skin. The fresh-air duct is usually made of some type of fibrous material, and the preheater duct is typically fabric-covered aluminum. If these ducts are missing, have loose connections, or have been damaged by punctures or by being squeezed shut, they could be causing your engine's poor performance.

The amount and temperature of the air fed to an engine is of extreme importance to its performance, economy, and durability. If a fresh-air duct is missing or damaged, the engine must rely on under-the-hood air to keep it going. Most engines don't like that, especially if it is a hot day and the resultant fuel and hot air mixture causes it to perform erratically.

Same thing with the preheater duct. If it is damaged in any way, the engine can't receive correct amounts of preheated air under cold operating conditions. And that's important during cold weather.

Without preheated air, the engine gets confused, cantankerous. It doesn't like running on cold air when it, too, is cold. It should have its fix of preheated air when it is warming up, and the preheater duct provides it.

Once the engine has warmed to the correct operating temperature, a thermostatically controlled valve in the air cleaner closes. This shuts off the supply of preheated air and allows fresh air to enter by way of the fresh-air duct.

Many older cars do not have carburetor air ducts. Look for special openings on your air cleaner to determine if it is one that should be fitted with ducts.

Take a few minutes and check those air ducts on your engine. If loose, they should be tightened; if damaged, they should be replaced; if absent, new ones should be installed. Give your engine what it craves—hot air when it is cold, and cool air when it is hot. It will repay you with more miles per gallon, better performance, reduced emissions, easier starting, less stalling, and no carburetor icing.

Air ducts are easy to replace and relatively inexpensive. They can be found at most auto-parts stores. Don't duct this issue; check your ducts now.

THE WHEELS

Wheel bearings that are too tight will create excessive friction and drag at the bearing/sleeve juncture and won't allow the wheels to turn freely. The adjusting nut must be torqued to specifications.

Improper front-end alignment can cost you an extra 1 to 2 gallons of gas per tankful. Bad alignment can cause the car to pull heavily to one side or the other, or it may force the front wheels to turn in and literally plow the road. A car with bad alignment will want to go one way while the wheels try to go another. This constant battle between forward movement and drag created by the misaligned wheels costs you more than just gasoline because tires (and steering components) wear much faster.

Dragging brakes are too often a factor in poor gas mileage. When the brake shoes constantly rub against the drum, unnecessary friction is built up at the wheels. The engine must then develop additional horsepower to overcome the friction-created drag. Suspended wheels should rotate easily and be free of any scraping sounds when spun by hand. **Note:** With disc brakes, there will probably be some sound, as friction pads are always lightly touching the revolving wheel.

An emergency or parking brake that is adjusted too tightly will cause the same condition as dragging foot brakes. Wheels should be completely free once the brake is released.

Many drivers forget that rear wheels can get out of alignment too. Although all cars don't require rear-wheel alignment, some do. Check your owner's manual or ask your dealer if yours is a car that does. If you notice the rear tires starting to wear on the inside, that's usually a sign that they are out of line. In some cases, rear-wheel alignment can be done only by a new-car dealer's service department.

EMISSIONS-CONTROL EQUIPMENT

All modern cars have a "check engine" or "service engine soon" light on the dash. If this light comes on, it usually indicates that some part of the emissions-control system needs attention. Don't ignore it. Check your owner's manual to see exactly what this warning means for your car. Emissions-control system malfunctions will negatively affect fuel economy.

On post-1966 automobiles, the entire emission-control system usually consisted of a simple PCV (positive crankcase ventilating) valve. This simple check valve meters recycled crankcase fumes and allows them to be drawn into the intake manifold, where they are reburned. If this valve becomes clogged or stuck, the air/fuel balance is upset and rough idling and poor performance result. Replace the PCV valve if you are in doubt.

A bad PCV valve affects fuel economy in a number of ways: The fuel/air ratio is altered, the ignition advance vacuum is upset, and friction is increased due to the accumulation of pollutants in the oil. In extreme cases, pressure builds up to the point where minor gaskets, such as valve-cover gaskets, may blow out.

Most service stations will check and/or replace a PCV valve for a small fee. It is usually located in the valve cover and can be checked by starting the engine, removing it from its housing, and shaking to see if it is free. It can also be checked by removing a hose from the valve cover end of the PCV, placing a finger over the end and checking for suction. If none is present or if the pull is weak, the valve needs replaced.

Most cars today use a closed rather than open PCV system. Fresh, filtered air is routed from the air cleaner through the crankcase, where it mixes with the crankcase fumes and is then metered back into the engine via the carburetor or intake manifold. It is then burned along with the contaminated air it picks up in the crankcase.

These systems usually work wonderfully well. But they do need care because if they are not operating properly, they can cause rough engine performance, a drop in fuel economy, hastened crankcase-oil deterioration, and accelerated deposit formation in critical areas of the system.

Remember, any tune-up, major or minor, should include a complete check, cleaning, and replacement (if necessary) of all parts of the PCV system.

On newer cars the EGR, or exhaust gas recirculation valve, should be checked to make certain it isn't stuck, because it controls the amount of burned gases from the exhaust system that are recirculated back into the engine. A frozen valve can result in a mileage drop and poor idling. A sudden dramatic drop in fuel economy can sometimes be traced to a stuck or inoperable EGR valve. Many cars have EGR warning lights on the dash that indicate when it needs servicing.

One engine sensor that should be checked periodically is the oxygen (O_2) sensor. The O_2 sensor is positioned somewhere in the exhaust stream of the engine and senses the amount of oxygen in the car's exhaust. It then relays this information to the car's computer which in turn adjusts the fuel/air mixture accordingly to get top performance and economy.

One survey has shown that of all the emissions control components on a vehicle, the O_2 sensor is the most likely to develop problems. In a test conducted on a group of California fuel-injected cars that failed the emissions test, 68 percent of the failed cars needed a new O_2 sensor.

Many cars have an O_2 sensor light that comes on when the O_2 sensor needs servicing or replacement. For top fuel economy and minimal emissions, the O_2 sensor must be operational. It should be changed or serviced according to the manufacturer's instructions.

Cars equipped with the SDS (spark delay system) pollution control have a spark delay valve in the exhaust-emission-control system. It is important that this valve be checked and replaced if it is not functioning. Otherwise, it may become a major cause of poor mileage because its action directly affects the distributor-advance mechanism.

Check your air cleaner to see if it has a small filter element where the crankcase vent hose attaches. If your car is so equipped, this filter must be cleaned and replaced periodically for proper operation of the emission-control system.

Don't tamper with any emission-control system in the hopes of getting better mileage. Besides being illegal in most states, it can do more harm than good. In an EPA test, private service garages were asked to try to get better mileage from a number of vehicles equipped with emission-control devices. How did they do? EPA noted that such tampering is more likely to hurt fuel economy than to improve it. Tampering virtually always makes emissions worse, and can cause deterioration in engine durability. Regular maintenance according to manufacturer specifications improves both emissions and fuel economy. Figure 6-6 shows the results of this test.

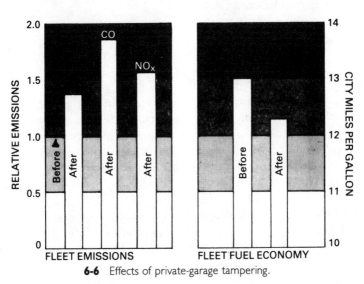

6-6 Effects of private-garage tampering.

Many states now require emissions testing of all vehicles over a certain age. All states should require it! Emissions reports give clues to as to your engine's health and operating efficiency. The higher the emissions, the less efficient the car is operating, and the less mile per gallon it will get.

Save your emissions reports, along with a record of your gas mileage. They are useful tools in helping you determine how the engine is doing and if tune-up or repair time is near.

Although it is not within the scope of this book to give a detailed explanation of computer controls and electronic sensors on modern car engines, a number of sensors have a direct bearing on fuel economy and require periodic inspection/replacement to ensure they are working properly. Although all engine sensors in some way might affect a vehicle's fuel economy potential, the following are among the most important.

- Coolant temperature sensor.
- Temperature sensor/switch.
- Throttle position sensor.
- EGR position sensor.
- Oxygen (O_2) sensor.
- Ported vacuum switch.
- EGR valve.
- PCV valve.

THE EXHAUST SYSTEM

Be certain that there are no restrictions in any part of the exhaust system, from the exhaust manifold to the end of the tail pipe. Check for a clogged, bent, or dented exhaust pipe, tail pipe, and muffler. Any restriction in the exhaust system will cause an increase in back pressure, and total performance will suffer.

Your tail pipe can tell you if you are wasting gasoline. A wet, black, sooty tail pipe that smells of gasoline (with no appreciable trace of oil) probably indicates a too-rich carburetor mix, bad or obstructed fuel injector(s), a stuck or maladjusted automatic choke, or an inoperable thermostat or manifold heat valve. Ideally the inside of the tail pipe should be dry and of a light gray/brown coloration. Any other condition means that the engine isn't burning fuel efficiently.

Catalytic converters won't last forever. A clogged or dented unit will adversely affect fuel economy and engine power. If you notice any change in your engine's operation and/or a sudden drop in fuel economy, check the catalytic converter for proper operation. A visual check can determine if there are any dents or holes that may affect its performance.

Sometimes a converter that is going bad will tell on itself. A slow, gradual loss of fuel economy could be an indication that the converter is slowly plugging up.

THE COOLING SYSTEM

An engine that runs too hot won't be as efficient and will wear out faster than one that operates at proper temperature. Such a simple thing as occasionally cleaning the debris (leaves, paper, bugs) from the front of the radiator will help it do the job it's meant to do. Keep the engine at normal operating temperature.

Another item that we don't usually associate with fuel economy is the radiator cap. If it doesn't seal properly the cooling system won't operate at its most efficient temperature, and it might take a lot longer to warm during the fuel-inefficient, cold-start period. The cooling system may never fully pressurize and reach its correct operating temperature. A cap in good condition ensures the engine will heat rapidly and hold its temperature. If your cap is damaged, or the gasket is worn or frayed, replace it.

New caps aren't expensive; if yours is questionable or you are not sure it is the proper pressure range, check your owner's manual. A car dealer's service department, auto-parts store or gasoline service station should also have the correct information or a new cap.

THE BODY

If two cars are exactly alike in every respect, but one is carrying four passengers while the other has only the driver, the one
with the extra passengers will suffer a 6-12 percent decrease in
gas mileage due to the weight of the extra occupants. Weight
has a direct bearing on gas mileage. For every 100 pounds of
added weight, mileage decreases from 1-6 percent, depending
on the size of the car. Check the trunk and the rear seat, and
remove any unwanted articles—it costs you to lug them
around.

Don't carry extra gasoline in your car. Besides adding unnecessary weight, it's a dangerous practice.

Figure 6-7 dramatizes the effect of weight versus fuel mileage.
It shows that additional weight has a more pronounced effect
(in terms of mileage) on a small car with a small engine than it
does on a large car with a large engine. But with either car it's
still you, the driver, who pays the penalty.

If you must carry extra weight, don't be penalized twice when
you carry it. Store things in the trunk or in the car's passenger

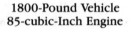

1800-Pound Vehicle   **5000-Pound Vehicle**
85-cubic-Inch Engine   **429-Cubic-Inch Engine**

With 100 Pounds Added   With 100 Pounds Added
Fuel Mileage Decreases   Fuel Mileage Decreases
2.1 MPG   0.28 MPG

6-7 Weight vs. engine size.

compartment rather than on the roof or the trunk rack where you will be further penalized (in addition to the extra weight penalty) in fuel used to overcome the wind resistance generated by the objects placed on the exterior of the car.

Speaking of weight, I don't have to tell you that it takes a lot of extra gasoline to tow another vehicle or trailer behind your car. Taking it easy when towing is always good advice. Don't try to go at the same speeds you would when not towing. By adjusting your speeds to what the engine can handle comfortably—this is where a vacuum gauge, tachometer, and trip mpg computer come in very handy (see chapter 4)—the mileage penalty you pay for towing won't be quite as severe.

It has been demonstrated that a 10 percent decrease in aerodynamic drag (the car's susceptibility to wind resistance) of an intermediate-size car results in a 5 percent increase in gas mileage at 60 mph. A cleaner, more aerodynamic profile means better gas mileage, and although you can't do much to alter the overall shape of your car (spoilers are an exception), you can clean up what you have. Luggage and ski racks should be removed when not in use. Pennants on antennas, bug deflectors, too many or oversized mirrors, big hood ornaments—yes, even mud flaps—create air turbulence and hold back the car. Clean up the outside of your car for improved mileage.

If you live in a state that requires a front license plate, position it so that it doesn't contribute to the car's aerodynamic drag. If it hangs down into the air stream, move it up and fasten it to the bumper.

Keeping your car washed and waxed is another effective way to lessen aerodynamic drag and gasoline consumption. A smoother surface will shed wind better, enabling the car to move with less power output.

The following was from the original edition of this book. I have left it unchanged to show how many of the ideas in the study noted have actually come into being in the past decade. Just about every suggestion (with perhaps the exception of the light-weight diesel engine) is now everyday reality on modern cars.

What's in store for drivers in the near future? A study conducted for the Department of Transportation and the EPA had as its goal a 43 percent improvement in fuel economy for all passenger vehicles by the year 1980. The study concludes:

"The most cost-effective approach—with medium technical risk . . . involves use of a light-weight diesel engine, combined with a four-speed automatic transmission, a torque converter equipped with a lock-up mounted in a light-weight improved body, and a chassis equipped with radial tires."

Chapter **7**

Regular automobile maintenance

Every moving part of your car has an effect on gas mileage! A car that is well lubricated delivers better mileage than one that isn't. Follow a pattern of regular-interval lubrication as recommended by the manufacturer—more often under extreme conditions. Ball joints, universal joints, differential, and other lube points work better when properly lubricated. Remember that power loss due to friction occurs at every bearing and gear in your car. Regular lubrication will keep the loss at a minimum and save you gas.

Front-wheel bearings should be cleaned and repacked with proper-grade lubricant (see Chapter 4 on moly and PTFE bearing lubricants) every 10,000 miles or at manufacturer's suggested intervals. Wheel bearings are a critical point at which the moving wheels meet the stationary car frame—an area of extreme friction buildup. They must be clean, well-lubricated, and free of burns or pits if friction is to be minimized. A bad bearing can usually be detected by a loud intermittent rumbling or grating coming from one of the front wheels. It should be replaced immediately.

Clean the oil breather cap (the one you remove to add oil) at lease once every 5,000 miles, and more often under severe dust conditions. A clogged cap affects the engine's air-drawing ability, and if it is not clean, will cause it to use more gas. Some cars have two, so be sure to clean both. Parts cleaner, kerosene, or gasoline will do the job.

An engine that is clean on both the outside and the inside goes hand in hand with better mileage because a clean engine runs better and is less likely to develop problems.

Clean oil = clean internal engine = better performance. It is essential that the engine oil and the filter be clean. Change both the oil and filter at prescribed intervals, preferably every 3,000 miles or every three months, whichever comes first—again, more often if weather conditions dictate. Oil must be clean to be able to reach and lubricate vital engine parts. Dirty oil carries in it the tidings of early engine failure. Oil is the life-blood of the engine, so don't skimp when it comes time for a change. Invest in a high-quality, reputable brand; it's worth the few extra nickels per quart, and is excellent insurance against future engine problems.

When transmission or axle fluids get old, they accumulate a lot of dirt and other particulate matter. This increases the overall viscosity of these fluids and requires the car to use more power to run the units because the gears must work in thicker, dirty fluids. Change these fluids according to the manufacturer's instructions and the gears will work in clean, correct viscosity fluids—not thick, dirty, abrasive, and contaminated old fluids. You will add life to the mechanical parts and increase fuel economy.

Maintain engine oil at the correct level. The engine needs all the friction-fighting protection it can get. Low oil level won't provide adequate lubrication, and engine efficiency will drop. Keep the oil full.

Oil also acts as the primary cooling fluid for the lower part of the engine. The more oil there is (up to the full mark on the dipstick), the better it can cool.

The same applies to the transmission fluid. Check it often and keep it full for best performance. In fact, *all* fluid reservoirs should be checked and topped off regularly to ensure they will operate at maximum efficiency. I have mentioned oil and transmission fluid levels as two of the most important, but don't ignore the rear-axle fluid, brake fluid, power-steering fluid, radiator coolant, and battery fluid levels.

The *heat riser* is a thermostatic valve located at some point along the exhaust manifold. It must move freely to ensure maximum performance from your car. During engine warm-ups, the valve remains closed, forcing warm exhaust gases to stay a bit longer in the manifold and hasten the warm-up. Then, as the engine warms, the heat riser opens and permits normal passage of exhaust gases. A sticking or frozen heat riser can cause very rough idling and poor response from a cold engine. Spraying the valve with penetrating oil while giving it a few taps with a hammer will usually free it. No tune-up should be considered complete unless the heat riser has been checked to be sure it is free.

The importance of regular-interval tune-ups cannot be overemphasized. If you take your car to a mechanic, be sure it is tuned at least once a year (more often for older, pre-computer cars), or a minimum of every 12,000 miles. Look for a shop that offers a dynamometer tune-up. This instrument can simulate actual road driving conditions and tune the car accordingly. The result is a more accurate and lasting tune-up. Even better are shops equipped with a modern computer engine analyzer.

Older car tip: If you do your own tune-ups, it's a good idea to work on the distributor, plugs, and timing first. After these are satisfactory, then do your carburetor adjustments. Why? Because ignition can affect the carburetion. For a better, truer tune-up, adjust the carburetor last.

The following section has been partially reprinted by permission of the New York Times Syndication Sales Corporation from Bob Sikorsky's "Drive It Forever" column.

Just about everyone knows that a car in need of a tune-up won't start as easily or give as good of performance or gas mileage as one properly tuned. According to Champion Spark

Plug Co., the condition of the ignition system is the single most important factor in starting your car in cold weather. Champion says it's even more important than the condition of the battery. That it's also critical for good fuel economy, especially in cold weather, goes without saying.

When an ignition system has broken or cracked ignition cables, worn spark plugs, or a worn or corroded distributor cap or rotor, starting the car, even with a new battery, can be a chore. Cold weather compounds the problems. Electricity, like water, seeks the easiest path along which to flow. Faulty wiring and bad ignition components interrupt or drain the electrical flow. Worn spark plugs, for example, easily require twice the electrical current as new ones.

Cold also affects the capacity of the battery to produce the necessary power for starting. At 80 degrees Fahrenheit, a battery has maximum power. That drops to 60 percent at 32 degrees, and goes down to 46 percent at 0 degrees.

Tests at Champion demonstrated the relationship between voltage required and voltage available. Engineers attempted to start a Chevy Camaro with a conventional ignition and a Dodge with electronic ignition at 0 degrees. Before the tests, both cars were able to start in warm-weather conditions. After being "soaked" in zero-degree cold the following was noted:

The Dodge, with its battery and engine in their original condition, did not start in four 30-second attempts. A new battery was installed, but another attempt at starting was unsuccessful.

Technicians then tuned the car, replacing old spark plugs with new ones, resetting the timing to factory specs, and replacing distributor components. Using the original battery, five new attempts average starts in 2.67 seconds.

In the Camaro tests, with the battery and engine still in as-is condition, there was one unsuccessful start, one start after 9.52 seconds and one start at 2.31 seconds. When a new battery was installed, the average starting time was 9.87 seconds in eight attempts. With new spark plugs and the original battery, the average starting time was 1.67 seconds. With the engine tuned and still using the old battery, starting time averaged 1.75 seconds.

A new battery could not help start these cars at zero degrees, yet once the engines were tuned even the old battery could provide sufficient voltage to fire the engine.

The conclusion wasn't surprising. Champion noted that a prewinter tune-up is indispensable for dependable starting. I might add that it's also indispensable for maximum wintertime fuel economy.

One of the main keys to maintaining your car is the owner's manual. When was the last time you looked at yours? Unfortunately, instead of being one of the most-read books, the owner's manual is one of the most ignored. If you want better economy from your car, it should be high on your list of priority reading. *Don't ignore the owner's manual.* It's a specific treatise written for your car and your car only. It's full of helpful information that will lead you down the road to more miles per gallon. Combined with this book, the owner's manual gives you a powerful one-two punch in the fight to get better mileage.

Modern cars are loaded with high-tech electronic equipment and computers that control most every engine function and some transmission and suspension operations. Fixing and maintaining a modern car is not a job for a novice; indeed, the days of the shadetree mechanic are all but gone.

There is very little even a mechanically inclined car owner can do to maintain, repair, or tune a newer car. A shop with updated equipment and personnel who are trained and can keep up with the advances in modern technology is a must for people who want maximum performance, fuel economy and longevity from their vehicles. My book, "Rip-Off Tip-Offs: Winning the Auto Repair Game," will help those not mechanically inclined or familiar with the repair scene, to find that honest, competent technician and shop that will guide them on their way to longer car life and more miles per gallon.

Chapter **8**

How tires can
save you gas

Stay away from the popular wide-track tires if you want top mileage from your car. They are strictly performance and looks tires, and they will cut your mileage. The narrower the tread width of a tire, the better its gas-conserving qualities. Narrow- or standard-width tread tires produce less friction at the road surface and create less rolling resistance, making it easier for the car to move. If it were possible to equip cars with bicycle tires, gas mileage would be phenomenal.

"Would it surprise you to discover that your tires are significantly underinflated? Chances are they are 8 pounds to 18 pounds less than what they should be. Of 100 cars we checked, only one was at the rated pressure printed on the side of the tire! Most were in the mid-20s, and quite a few were as low as 15 pounds. The record was a squat tire with only 8 pounds of air pressure."

So begins an Arizona Energy Office pamphlet on ways to save gasoline. Keeping tires at the tire manufacturer's recommended maximum air pressure is one of the easiest ways to save gasoline and increase tire life—and by far the cheapest.

The Arizona Energy Office pamphlet also notes that if your tires are normally inflated to 24 psi (pounds per square inch), and you increase the air pressure to a maximum of 32 psi, your gas mileage should increase by 3 miles per gallon. That can mean a savings of about $130 for someone who is now spending $800 annually on gasoline.

Remember to always check and adjust tire pressure when tires are cool. Adding an extra 3-5 pounds of air to each tire won't noticeably affect the riding qualities of your car but it will increase tire life and gas mileage by reducing the rolling resistance and friction generated by overly soft tires. Air is free, so don't be afraid to use a little more of it and enjoy an extra 1-2 miles per gallon.

Inflate tires to the maximum pressure printed on the side-wall of your tires. This number will usually be between 32 psi and 35 psi. Flabby tires are one of the most overlooked items on a car, and many surveys have shown that soft, underinflated tires are all too common of a phenomenon on our roads.

Be sure to add more air to the tires when the weather turns colder. A 10-degree drop in temperature will reduce tire pressure by 1 pound. The difference between summer and winter tire pressures could be as much as 8 pounds less. This could cost you 2 mpg if not corrected.

Check tire pressure often—once a week if possible, once every two weeks at the minimum. A slow leak of just a few pounds a week can silently rob you of gasoline and hasten tire wear.

If you don't know how to use a tire pressure gauge and an air hose, ask someone to show you. There is no easier preventive maintenance practice that pays such handsome dividends. Longer tire life, extended engine life (the engine doesn't have to work as hard to overcome the tire's rolling resistance), and increased fuel economy are some of the major benefits of keeping up with your tires.

Recent fuel-economy tests conducted by Firestone showed overwhelmingly that radial tires improve gasoline mileage when compared with bias-ply or nonradial tires. Improvements in gas mileage ranged from 7-10 percent, depending on the speed of the test cars. Translated into miles per gallon, this means that a car now getting 15 mpg with conventional tires would improve to 16-16.5 mpg simply by changing to radial tires. Projecting this further, you would use 205 less gallons of gasoline if you drove the radials a full 40,000 miles, the usual guaranteed mileage. At $1.50 per gallon, this means a savings of over $300 in gasoline alone—a very impressive figure. Be sure to consider radial ply tires when the time comes for replacement, and enjoy the added savings and safety they provide. If you do a significant amount of highway driving, radials can be especially important and beneficial to you.

Figure 8-1 demonstrates the superiority of radial tires over bias and bias-belted types. Rolling resistance, a major factor in gas mileage, is significantly less with radials.

Here's something to keep in mind about that new set of tires you just bought for your car. Rolling resistance of a new tire is higher than that of used tires. *Rolling resistance* is the friction or resistance generated by a tire as it moves. More rolling resistance means lower miles per gallon. Expect a slight decrease in fuel economy until your new set of tires breaks in.

Using oversized-diameter tires on the rear of your car has the same effect as a high-speed rear-axle ratio (for rear-wheel drive cars). By using larger tires on the rear, more distance is traveled with each revolution of the rear axle, and slightly better gas mileage is realized. You may have to contend with a slight power loss in lower gears.

Before you switch to bigger diameter tires on the rear of your car, check with the manufacturer to be certain the replacement tires you are considering are safe replacements for the size tire that came with the car. Some overly large tires could adversely affect the handling of the vehicle.

Although it would be a little impractical, if you really wanted top mileage and would go to extremes to get it, driving on treadless tires would give you a big head start. Look at FIG. 8-2 to see the effect tread removal has on rolling resistance. It cuts it in half! So if you're ever in a mileage marathon, "baldies" will up your gas mileage considerably.

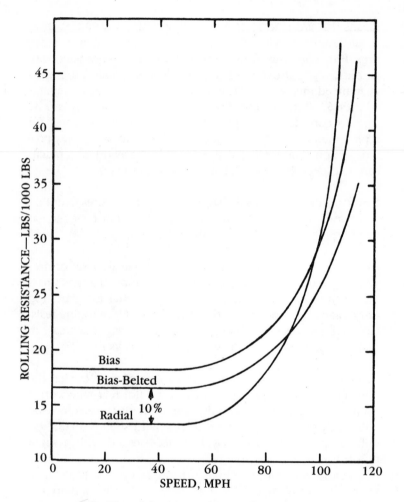

8-1 Effect of tire construction on rolling resistance.

If you live in a rainy climate, it will pay you to shop for tires with extra-wide grooves in the tread. All-weather or rain tires lessen rolling resistance caused by water buildup on the road by channeling away the water more efficiently. They are definitely a mileage plus in rainy areas.

Tires that are not balanced properly tend to hop and shimmy, creating additional rolling resistance and drag that the car must

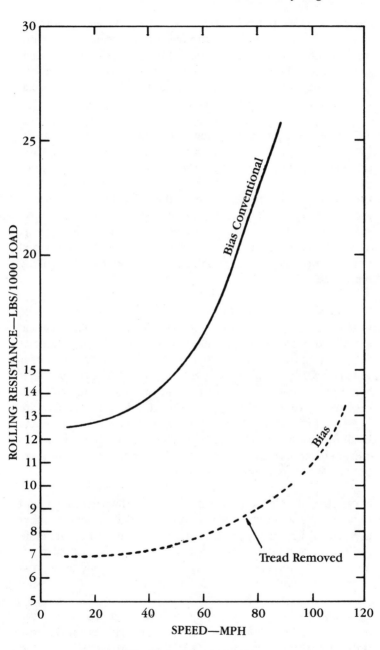

8-2 Effect of tread removal.

overcome. Well-balanced tires are a prerequisite to top gas mileage. Be sure to balance the rear tires as well as the front, because they tend to exhibit similar characteristics when unbalanced. Then, when rotating your tires, you won't have to bother balancing the rear ones. If the spare is used in the rotation, it should be balanced also.

Whenever you purchase a new set of tires, don't have them balanced right away. New tires should be driven a few hundred miles so that they can acquire a set, each tire conforming to its own particular traits. If no balance weights are present to affect the tire during the break-in period, the set will be truer. Drive a few hundred miles on the unbalanced tires, and then have them balanced—it will be more accurate and longer lasting. You will gain many extra miles of tire wear, and economy will improve because the tires generate considerably less rolling resistance.

Sometimes it is practically impossible to get a tire to balance because of an imperfection in its basic structure. This is occasionally the case with so-called *blems*. Too much rubber at points along the tread surface make the tire uneven and a bear to balance. If you are stuck with this type of tire, try having it trued. Tire *trueing* is available in most larger cities and will probably cure the existing problem. Thin layers of rubber are peeled off the tire until the surface is even and the radius equal in all directions. It could save you buying another set of tires and will improve mileage to boot. For more information on tire and wheel alignment, see Chapter 6.

Snow tires should be installed no sooner than necessary and removed as soon as weather permits. The deep cleats on snow tires are designed for traction in snow or mud and generate more friction and rolling resistance than conventional tires. You pay a 1-3 mpg penalty when driving with snow tires, so as soon as they aren't needed, take them off.

Get those snow chains off the car—use them only under the severest weather conditions. Driving with chains on the rear tires is like driving with an extra 1,500 pounds of weight in the car.

Chapter **9**

Accessories that hurt gas mileage

Any time you use a battery-powered accessory, you use extra gas. Radio, tape deck, cigarette lighter, power seats, power windows, interior lights, heater and blower fan—all use electrical current. When one of the accessories is used, the alternator/generator is activated to restore to the battery what is being drawn off. This extra work of the generating unit is paid for with gas burned to provide the additional horsepower needed to turn the unit. Use accessories sparingly. Remember that accessories cost you gas in two ways: It takes extra horsepower to run them; and their bulk adds considerable weight to the car, and you must pay with your gasoline dollars to haul that weight around.

Use of battery-operated accessories while the engine is off must be paid for with gas burned once the engine is started again. The alternator must go to work again, charging the battery. When the ignition is off, be certain lights, radio, and all other battery-operated accessories are also off.

Roll up power windows before stopping the car. This prevents running the engine while waiting for the windows to roll up— a small item, to be sure, but it saves a few seconds' idling time.

Sun roofs are popular new car options. However, driving with a sun roof open will cost you miles per gallon because of the extra wind resistance it causes. It acts just like an open window. Use the sun roof sparingly.

Turn off the air conditioner whenever you park the car. This will save gas when the car is restarted because the engine won't have the extra drag of the compressor lowering cranking efficiency.

If you don't think you pay a gas-mileage penalty for running your air conditioner, just glance at the comparative gas mileages of the car with A/C on, and A/C off in FIG. 9-1. As you can see, the penalty may run as high as 4 mpg at speeds of 20-40 mph. Be selective when using the air conditioner; use the "vent" position when full cooling is not required. Save the air conditioner for the really hot days, and you can gain up to 4 extra mpg while it is off.

Turbochargers or turbos have become commonplace on many of today's new cars. Turbos are offered as either options or

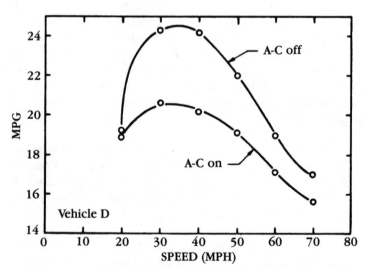

9-1 Road-load fuel economy.

included as part of a "performance" package. A turbo uses the car's own exhaust gases to drive a turbine wheel(s), which in turn forces outside air into the engine. The additional concentrated air increases the engine's horsepower.

Will turbos cost you mpg? Probably. If you have a turbo you are likely to use it, and most turbos don't activate until the accelerator pedal is significantly depressed. That means that every time you activate the turbo, you pay for it in extra gasoline used.

You can get good gas mileage with a turbocharged car if you are selective and only call on the turbo's boost for passing or emergencies. Any way you look at it, a turbocharged vehicle requires more attention if you want high mileage than does a naturally aspirated one.

In recent years, sports-utility vehicles have become popular with the American public. These go-anywhere do-anything rough riders have been a boon to people seeking recreation or just wanting an all-purpose vehicle that is capable of in-town luxury in an off-road environment.

Most sports-utility vehicles are equipped with either full-time or part-time 4-wheel drive. Full-time 4-wheel drive will get poorer fuel economy than part-time 4-wheel drive, where the driver can select when he wants the vehicle in 4-WD mode. Vehicles equipped with part-time 4-WD will get less fuel economy than a comparable 2-WD vehicle.

Four-wheel uses more fuel because the engine has another axle to power. In addition, these vehicles are usually geared lower to allow for better pulling power over rough roads. Combine that with the extra weight of the extra drive axle and the fact that these vehicles usually use wide, all-terrain tires, and you have a formula for fuel inefficiency.

If you are contemplating a 4-wheel drive, shop carefully and check the vehicle's EPA estimated fuel economy. Some 4-WDs are more efficient than others. Look at the EPA fuel economy estimate charts in Chapter 12 to see how widely the fuel economy of current 4-WDs vary.

Chapter **10**

Other conditions
that cost you mpg

The type of road surface you drive on has a direct relation to the mileage you get. A loose-gravel surface can cut mileage by a full third, a muddy road even more. The smoother and firmer the road surface, the better your mileage will be. Stay with good roads when you can.

Some conditions that can influence fuel economy are listed in FIG. 10-1. Their economy penalties are based on steady cruising at about 50 mph.

Remember that the *berm* of the road—that is, the angle at which each side of the road slopes away from the center—may differ from state to state. If, upon changing locales, you notice that your car now pulls to one side, have it checked for alignment. It may have to be reset to conform to the new roads you are driving.

Avoid driving on wet roads unless it is absolutely necessary. The car must put forth extra effort to push itself through the water on the road, and you can lose 1 mpg on wet surfaces. Government researchers on fuel economy recognize this fact and terminate any economy tests "if the pavement becomes damp enough that the car leaves visible tracks."

Road Conditions:	**MPG loss**
Broken & patched asphalt	15%
Gravel	35%
Dry sand	45%
3% Grade	32%
7% Grade	55%

Environment:	
18 MPH tailwind	(19% gain)
18 MPH crosswing	2%
18 MPH headwind	17%
50°F ambient temperature	5%
20°F ambient temperature	11%
Altitude (4000 ft)	15%

10-1 Effects of road conditions and environment.

10-2 Fuel consumption vs. air temperature.

If rain on the road can cost you 1 mpg, snow is even worse. As little as 1 inch of snow on the road is a formidable barrier for the car. It's like constantly climbing a hill, and that requires a lot of extra gas. If you must drive on snow-covered roads, be prepared to have your mileage drop considerably.

Scrape all snow and ice off the car before starting out. Wet snow can really weigh a car down. Don't use extra gas lugging it around.

The colder the *ambient* or outside temperature, the worse your mileage will be. This is due to the fact that cold air is more dense and increases aerodynamic drag. Reduce your top speed a bit during winter months and you will offset any mileage loss caused by the colder air. Figure 10-2 gives fuel consumption versus air-temperature curves for a car traveling at 50 mph.

Strong wind will have either a positive or negative effect on gasoline consumption, depending on the direction from which it comes. To illustrate: U.S. Department of Transportation studies have shown that a full-size test car got 13.9 mpg at

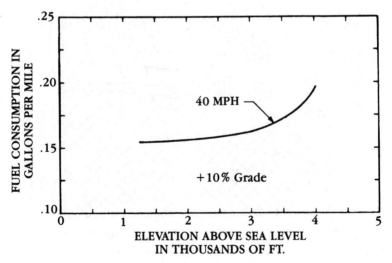

10-3 Fuel consumption vs. elevation.

70 mph with no wind. The same car, traveling at the same speed but with an 18 mph *headwind*, got only 11.6 mpg. A car going 70 mph with an 18 mph *tailwind*, got 16.6 miles/per/gallon—a whopping 5 mpg increase over the car that had to buck the headwind.

Use the wind to your advantage. Strong headwinds should be a signal to slow down so you can reduce the effect of the extra wind resistance on the car. Don't fight headwinds.

A tailwind is a sign to increase your speed slightly and take advantage of the additional push the wind is providing. You'll

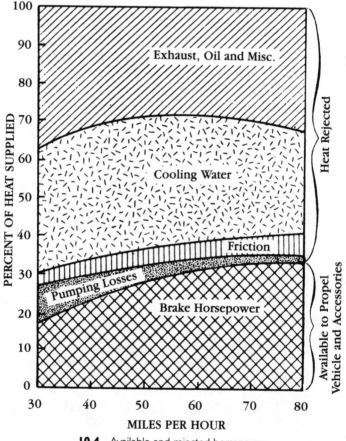

10-4 Available and rejected horsepower.

be able to go a bit faster and won't have to pay a gas penalty for doing so. Let the wind do some of the work, and ease up on the accelerator. Keep an eye out for changing wind directions and monitor your driving accordingly. Swaying trees or bushes, paper blowing across the road, smoke in the air—all give clues to wind direction and velocity. Use the wind when you can, it's free.

Figure 10-3 reminds the reader to expect a drop in gas mileage when driving at altitudes above 2,000 feet. The mileage penalty becomes even more severe when elevations above 4,000 feet are encountered.

Figure 10-4 is a sad commentary on automobile engine efficiency. Of all the energy released as heat when fuel is burned, only 20-30 percent is actually used to propel the vehicle at speeds above 30 mph. A full 70-80 percent is rejected heat and energy consumed by engine friction, accessories, and auxiliaries attached to the engine. No wonder it's a battle to get good mileage!

Chapter 11

"Gas-saving" devices
—Caveat Emptor

The recent rise in gasoline prices due to the Persian Gulf crisis has triggered another recent rise. This one is in the number of so-called "gas-saving" devices that have hit the market since the prices at the pump have escalated.

A word of caution: Almost every one of these "gas savers" is useless. There are no miracle cures for the gas mileage blues.

Remember: The money wasted on one of these devices is better spent on gasoline to run your car. You'll go a lot farther on the gas you buy with the money you'll save than you will on a tank of wishful thinking.

Perhaps the last paragraph of a letter I received from the Environmental Protection Agency (EPA) commenting on fuel-saving and emission-improving products best sums up the attitude any consumer should have when considering a device that promises too-good-to-be-true fuel, emissions or money savings:

"We strongly encourage any person who is considering purchasing any fuel savings or emission-improving product to first obtain data substantiating the claims. Additionally, we would strongly discourage a person from relying on testimonials in lieu of data generated from a competent independent test facility using appropriate procedures."

The EPA has published a list of the devices it has tested or otherwise evaluated. The EPA says that except for the ones noted otherwise, "none is expected to cause a statistically significant increase in fuel economy for a modern light-duty motor vehicle in proper operating condition which is operated in a typical manner." The list is reprinted below.

AIR BLEED DEVICES

Landrum Retrofit Air Bleed
Pollution Master Air Bleed
ADAKS Vacuum Breaker Air Bleed
Berg Air Bleed
Econo Needle Air Bleed
Monocar HC Control Air Bleed
Air-Jet Air Bleed
Aquablast Wyman Valve Air Bleed
Peterman Air Bleed
Mini Turbocharger Air Bleed
Ball-Matic Air Bleed
Landrum Mini-Carb
Econo-Jet Air Bleed Idle Screws
Turbo-Dyne G.R. Valve
Auto-Miser
Ram-Jet
Fuel Max**
Brisko PCV
Cyclone-Z

VAPOR BLEED DEVICES

Frantz Vapor Injection System
Turbo Vapor Injection System
SCATPAC Vacuum Vapor Induction System
Econo-Mist Vacuum Vapor Injection System
Mark II Vapor Injection System
Platinum Gasaver
V-70 Vapor Injector
Atomized Vapor Injector
Hydro-Vac
POWERFUeL

LIQUID INJECTION

Goodman Engine System, Model 1800
Waag-Injection System**

IGNITION DEVICES

Paser Magnum/Paser 500/Paser 500 HEI
BIAP Electronic Ignition Unit
Magna Flash Ignition Control System
Special Formula Ignition Advance Springs
Autosaver
Baur Condenser
Fuel Economizer

FUEL LINE DEVICES (heaters or coolers)

FuelXpander
Gas Meiser I
Greer Fuel Preheater
Jacona Fuel System
Russell Fuelmiser
Optimizer

FUEL LINE DEVICES (others)

Super-Mag Fuel Extender
Moleculetor
Wickliff Polarizer
POLARION-X
PETRO-MIZER
Malpassi Filter King

MIXTURE ENHANCERS (under the carb)

Energy Gas Saver
Hydro-Catalyst Pre-Combustion Catalyst System
Environmental Fuel Saver
Glynn-50
Sav-A-Mile
Turbo-Carb
Spritzer
PETROMIZER SYSTEM

Gas Saving and Emission Control Improvement Device
Turbocarb

MIXTURE ENHANCERS (others)

Electro-Dyne Superchoke
Filtron Urethane Foam Filter
Lamkin Fuel Metering Device
Smith Power and Deceleration Governor
Basko Enginecoat
Dresser Economizer

INTERNAL ENGINE MODIFICATIONS

ACDS Automotive Cylinder Deactivation System**
Dresser Economizer
MSU Cylinder Deactivation**

FUELS AND FUEL ADDITIVES

Stargas Fuel Additive
Sta-Power Fuel Additive
Technol G Fuel Additive
Johnson Fuel Additive
Vareb 10 Fuel Additive
Rolfite Upgrade Fuel Additive
QEI 400 Fuel Additive
EI-5 Fuel Additive
NRG #1 Fuel Additive
XRG #1 Fuel Additive
ULX-15/ULX-15D
SYNeRGy-1
Bycosin

OILS AND OIL ADDITIVES

Analube Synthetic Lubricant
Tephguard

ACCESSORY DRIVE MODIFIERS

Pass Master Vehicle Air Conditioner*
P.A.S.S. KIT*

Morse Constant Speed Accessory Drive* (not presently offered for sale in the aftermarket)

DRIVING HABIT MODIFIERS

AUTOTHERM*
Gastell
Fuel Conservation Device
IDALERT*

MISCELLANEOUS

Lee Exhaust and Fuel Gasification EGR
Treis Emulsifier
BRAKE-EZ
Fuel Maximiser
Kat's Engine Heater
Dynamix
Kamei Spoilers*
Gyroscopic Wheel Cover
P.S.C.U. 01 Device
Mesco Moisture Extraction System

* Indicated a statistically significant improvement in fuel economy without an increase in exhaust emissions, although cost-effectiveness must be determined by the consumer for his particular application.

** Indicated a statistically significant improvement in fuel economy but with an increase in exhaust emissions. According to Federal Regulation, installation of this device could be considered tampering.

Chapter **12**

Alternate transportation

Alternate transportation can provide you with unlimited miles per gallon. It's obvious that you can save gas by joining a car pool. You'll cut your gas bill by the number of people in the pool, so if you are not already pooling, try to find one you can join. Many cities offer a free service of computer matching for people who want to join a car pool. The places you live and work and the times you leave are matched with others in your area, and compatible pools are arranged. Check to see if your city offers a car-pool assistance program. The American Petroleum Institute says that the U.S. could save upwards of 33 million gallons of gasoline per day if we could add just one person to each individual commuter's passenger load.

A study by the Federal Highway Administration shows that if you drive 10 miles each way to work in an average-sized car, the yearly cost for driving is nearly $1,000. If you decide to join a five-man car pool instead of driving alone, you can save about $650 of that total. This money-saved figure includes gasoline, parking, insurance, and repairs. That's quite a saving! Find out if your company has a van pool. This is an efficient, entertaining, and economical way to get to your job.

Many large cities have special freeway lanes that are reserved for ride-sharing vehicles and buses during peak rush hour traffic. One of the saddest sights is to see these "express lanes" virtually empty. Most drivers ignore these time-, fuel-, and car-saving lanes, opting instead to fight congestion and gridlock day after day after day. Instead of spending that precious time with their families, they choose to sit in their cars, wasting gasoline, wearing their cars out, and wearing themselves out. Pretty dumb when you think about it.

Take advantage of those express lanes. Car pool, and you'll get home faster, save fuel, save your car and yourself, and maybe even make a friend or two in the process.

Although it's not my position to tell you where to go on vacation, it should be patently obvious that if you vacation closer to your home you will use less fuel than if you were to vacation far away. The money you save on gasoline can be spent on other vacation treats.

Most likely, your car is your primary means of transportation to and from work. But that doesn't mean you have to use it every day of the week. As we have seen, car pooling and van pooling are of great help. Buy why not set aside one day a week as public transportation day? Ride the bus, trolley, or train or use other modes of public transport to get to and from work or play.

The following section has been reprinted by permission of the New York Times Syndication Sales Corporation from Bob Sikorsky's "Drive It Forever" column.

I never like it when people preach to me. When someone gets on a soapbox and starts shoving his views down my craw, I usually retreat to some safe corner where I can effectively ignore his or her persuasions. I hope this doesn't fall into the preaching category, for it does attempt to persuade us to do something we all know we should be doing, but for some reason, don't.

She would never think of herself as a conservationist or an environmentalist, but when one looks at the way she has lived, she may indeed fill those shoes perfectly. Nor would she think that her mode of transportation throughout her 79 years had anything to do with helping keep our air clean or conserving fuel—or, for that matter, that is was anything out of the ordinary. It was what people like her did, out of choice rather than necessity, and never thought much about it. It was how one got from point A to point B. You see, my Aunt Stella was, and still is, the quintessential public transportation user.

Aunt Stella only owned one or two cars in her lifetime. I vaguely remember standing on the running board of her old Chevy when I was about 5 years old. A bit clearer is a 1951 Dodge. But since the Dodge, buses and trolleys, with an occasional train sprinkled here and there, provided transportation. Whether across downtown Pittsburgh, or south to Florida in the winter, or east to Atlantic City in the summer, public transportation provided the way.

But why this diatribe on my favorite aunt? Well, because if each of us driving a car had a little bit of Aunt Stella in us, the air would be a lot cleaner, our oil import quotas would drop significantly, and our roads would be a lot less congested. When they dole out the awards for conservation or environmentalism, my Aunt Stella and the many like her should be at the front of the line. Indeed, if the big oil and tire companies had to depend on Aunt Stella, they would have long since become dinosaurs.

The time is right for each of us to make a serious effort to try car or van pooling or to use some form of public transportation on a regular basis. I know that everyone agrees that this is a good idea, and most everyone I talk to says so. Public transportation officials have expounded the merits of alternate means of transportation for years and have made it easy for anyone wishing to participate. Car-pool matching, van pools, and public transportation are at anyone's fingertips in large cities. And walking—horror of horrors—is free to anyone who chooses to use it.

Motor vehicles are the number-one violator of our nation's air and the major user of oil. We could start weaning ourselves by leaving our vehicle at home just one day a week and using public transportation instead. Or if we can match up with just one more person instead of driving to work alone, that will cut in half the amount of gas we use and the pollution our vehicle spews into the air—a large payoff for a small effort. So let's try to imitate my Aunt Stella—at least once a week, anyhow, for starters. For if there were just a little bit more of Aunt Stella in each of us, we could take a lot of the unclean out of our air, preserve our finite resources, and cut our dependence on foreign oil.

The following section has been reprinted by permission of the New York Times Syndication Sales Corporation from Bob Sikorsky's "Drive It Forever" column.

When was the last time you rode a bicycle? That long ago? OK, how long has it been since you owned one? Even longer, you say.

What brings all this to mind is an incident that occurred recently when I was driving to town. A bicyclist was going in the same direction as I was, keeping as far to the right as possible, hugging the curb as best he could. A driver a few car lengths ahead of me wasn't satisfied with the bicyclist's demeanor, however, obviously feeling that the bike had no right to slow him down (these roads are for cars, you know!).

The driver took out his pent-up aggression on the slow-moving bicycle, literally running the pedaler off the road, sending him careening over the curb and almost crashing into a nearby wall. The bicyclist raised a clenched fist as the car whizzed on. The driver of the car, grinning, countered with a raised finger.

Bicycling friends of mine say incidents like this are fairly common, and those who rely on a bicycle as their main means of transportation say they look at every driver as a potential assassin, an adversary in a battle they can't possibly win. All have had close calls; caution is the byword at all times. On streets where there is no bike path, they must join the mainstream of traffic, and most do so with great trepidation. Many use side streets whenever possible to avoid heavy traffic. They are all keenly aware that in a one-on-one confrontation with a car, they don't stand a chance.

Why all this talk about bicycles in an automotive column? Mainly because the number of bicycles increases each year, and more and more of them are sharing the roads with cars and trucks. Although most bicyclists use designated bike paths whenever possible, the paucity of these protected areas forces them to use main traffic arteries at times. These are times when we, the pilots of those moving boxes of plastic and steel, should go out of our way to be extra careful when approaching a bicyclist.

I think every driver on the road owes something to every bicyclist. Why is that? Because bicycles aren't polluting the atmosphere with carbon monoxide, particulate matter, unburned hydrocarbons, lead, nitrous oxides, poisons, ad nauseam. Because they aren't adding to our tax burden by beating up on our roads. Because each one on the road makes your own town a much better place to live and breathe.

Bicycles are the ultimate mileage machine: unlimited mileage on zero gallons of gasoline, zero quarts of oil; quiet, clean, relatively fast, non-polluting, good for you and me. And if you own both a car and a bike, each time you use your bike you extend the effective life of your car. Why would anyone want to be aggressive with such a benign creature?

Oh sure, occasionally you may become a bit irritated when a bike is in front of you and you have to slow down, but that is a small price to pay for what we are getting in return. A bicycle benefits everyone; a car, just its owner.

So the next time you encounter a bicycle, slow down, smile, be extra careful, extra courteous. Remember, the gas it's saving could be yours; the air it's not polluting is everyone's.

Using public or other alternate transportation and car or van pooling may be the ultimate preventive maintenance practices. Every time you don't drive, your vehicle can't pollute. You save gasoline and money, ease traffic congestion, add miles to the other end of your odometer reading by preserving your car, and most importantly, you practice preventive maintenance on the planet itself.

For more information about better gas mileage, solid lubricants, and other items mentioned in this book, send a self-addressed, stamped, business-sized envelope to:

> THE MILEAGE COMPANY
> Box 40063
> Tucson, AZ 85717

Chapter **13**

The gas mileage guide

To aid you in your selection of a fuel efficient vehicle, this chapter includes a reprint of the most recent EPA gas mileage guide. You can obtain updated copies from any new car dealer or by writing to

> *Gas Mileage Guide*
> Pueblo, CO 81009

A guide can also be ordered by calling the National Highway Traffic Safety Administration's hotline number: (800) 424-9393.

The Gas Mileage Guide is published by the U.S. Department of Energy as an aid to consumers considering the purchase of a new vehicle. The Guide lists estimates of miles per gallon (mpg) for each vehicle available for the new model year. These estimates are provided by the U.S. Environmental Protection Agency in compliance with Federal Law.

The Guide is intended to help you compare the fuel economy of similarly sized cars, light-duty trucks and special purpose vehicles. The vehicles listed in the Guide have been divided into three classes of cars (sedans, two seaters, and station wagons), three classes of light-duty trucks (vans, small pickups, and large pickups), and three classes of special purpose vehicles (2-wheel drive, 4-wheel drive, and cab chassis).

By using this Guide you can estimate the average yearly fuel cost for the vehicle you choose to purchase. The mileage figures included in this Guide are most useful when comparing vehicles. Your actual mileage when driving a vehicle may differ considerably from the predicted mileage.

All new car and light-duty truck dealers are required to have copies of this Guide available and prominently displayed in their showrooms. The Department of Transportation is empowered to penalize dealers who fail to display the Guides as prescribed by law.

INTERIOR VOLUME

The interior volume is listed in the index of Guide for each body style (2-door, 4-door, and hatchback), except for two seaters and trucks, and is a way of estimating the space in a car. The interior volume is given as two numbers in cubic feet (for example: 87/12). The first number is an estimate of the size of the passenger compartment. This number is based on four measurements—head room, shoulder room, hip room, and leg room—for both the front and rear seats. The second number is the size of the trunk, or in station wagons and hatchbacks, the cargo space behind the second seat.

HOW THE FUEL ECONOMY ESTIMATES ARE OBTAINED

The fuel economy estimates are based on results of tests required by the U.S. Environmental Protection Agency (EPA). These tests are used to certify that vehicles meet the Federal emissions and fuel economy standards. Manufacturers test pre-production prototypes of the new car models and submit the test results to EPA. EPA then confirms the accuracy of the figures provided by the manufacturers. The vehicles are driven by a professional driver under controlled laboratory conditions, on an instrument similar to a treadmill. These procedures ensure that each vehicle is tested under identical conditions; therefore, the results can be compared with confidence.

There are two different fuel economy estimates for each vehicle in this Guide, one for city driving and one for highway driving. To generate these two estimates, separate tests are used to represent typical everyday driving in a city and in a rural setting.

The test used to determine the city fuel economy estimate simulates a 7.5 mile, stop-and-go trip with an average speed of 20 mph. The trip takes 23 minutes and has 18 stops. About 18 percent of the time is spent idling, as in waiting at traffic lights or in rush hour traffic. Two kinds of engine starts are used—the cold start, which is similar to starting a car in the morning after it has been parked all night, and the hot start, similar to restarting a vehicle after it has been warmed up, driven, and stopped for a short time.

The test to determine the "highway" fuel economy estimate represents a mixture of "non-city" driving. Segments corresponding to different kinds of rural roads and interstate highways are included. The test simulates a 10 mile trip and averages 48 mph. The test is run from a hot start and has very little idling time and no stops (except at the end of the test).

NOTE: To make the numbers in the Gas Mileage Guide more useful for consumers, EPA adjusts these laboratory test results to account for the difference between controlled laboratory conditions and actual driving on the road. The laboratory fuel economy results are adjusted downward to arrive at the estimates herein and on the labels seen on new cars (FIG. 13-1). The city estimate is lowered by 10 percent and the highway estimate by 22 percent from the laboratory test results. Experience has proven that these adjustments make the mileage estimates in this Guide correspond more closely to the actual fuel economy realized by the average driver.

Sample Fuel Economy Label
(Attached to New Auto Window)

This is the average estimate for city driving

Use these two estimates to compare to other models

This is the average estimate for highway driving

These numbers represent the range of fuel economy for other models of this size class

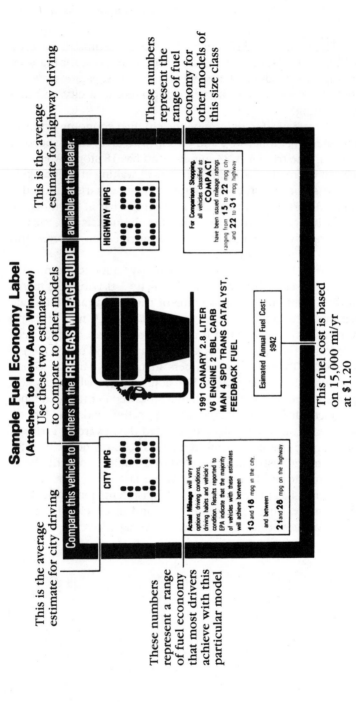

Compare this vehicle to others in the **FREE GAS MILEAGE GUIDE** available at the dealer.

CITY MPG

HIGHWAY MPG

1991 CANARY 2.8 LITER
V6 ENGINE 2 BBL CARB
MAN 4 SPD TRANS CATALYST,
FEEDBACK FUEL

Actual Mileage will vary with options, driving conditions, driving habits and vehicle's condition. Results reported to EPA indicate that the majority of vehicles with these estimates will achieve between

13 and **18** mpg in the city,

and between

21 and **28** mpg on the highway.

For Comparison Shopping, all vehicles classified as **COMPACT** have been issued mileage ratings ranging from **15** to **22** mpg city and **22** to **31** mpg highway

Esimated Annual Fuel Cost:
$942

These numbers represent a range of fuel economy that most drivers achieve with this particular model

This fuel cost is based on 15,000 mi/yr at $1.20

13-1 Sample fuel economy label.

How To Use This Guide

Compact Cars
(see vehicle classes below)

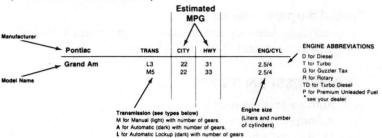

VEHICLE CLASSES

Sedans

Minicompact Less than 85 cubic feet of passenger and luggage volume

Subcompact Between 85 and 99 cubic feet of passenger and luggage volume

Compact Between 100 and 109 cubic feet of passenger and luggage volume

Mid-size Between 110 and 119 cubic feet of passenger and luggage volume

Large 120 or more cubic feet of passenger and luggage volume

Two-seaters Car designed to seat primarily two adults

Station wagons

Small Less than 130 cubic feet of passenger and cargo volume

Mid-size Between 130 and 159 cubic feet of passenger and cargo volume

Large 160 or more cubic feet of passenger and cargo volume

Trucks

Vans Passenger

Small pickups Trucks having Gross Vehicle Weight Ratings (GVWR, truck weight plus carrying capacity) under 4,500

pounds; 2-wheel Drive (2WD), 4-wheel Drive (4WD)
Large pickups Trucks having GVWR's of 4,500 to 8,500 pounds; 2-wheel Drive, 4-wheel Drive

Special purpose vehicles

All other light vehicles not in another car or truck class; 2-Wheel Drive, 4-Wheel Drive, Cab chassis.

TRANSMISSION TYPES

A3 Automatic three speed
A4 Automatic four speed
A8 Dual range automatic four speed
AV Continuously variable transmission
L3 Automatic lockup three speed
L4 Automatic lockup four speed
L8 Dual range automatic lockup four speed
M3 Manual three speed
M4 Manual four speed
M4C Manual four speed with creeper first gear
M5 Manual five speed
M5C Manual five speed with creeper first gear

GAS GUZZLER TAX

The Energy Tax Act of 1978 established a Gas Guzzler Tax on the sale of new-model-year vehicles whose fuel economy fails to meet certain statutory levels. The fuel economy figures used to determine the Gas Guzzler Tax are different from the fuel economy values found on the following pages. The tax does not depend on your actual on-the-road mpg, which may be more or less than the EPA published value.

The purpose of the Gas Guzzler Tax is to discourage the production and purchase of fuel inefficient vehicles. The amount of any applicable Gas Guzzler Tax paid by the manufacturer will be disclosed on the automobile's fuel economy label.

Check the Fuel Economy Label on the vehicle at the dealer's showroom for its mpg ratings. The mpg will vary because of engine emission controls and fuel system differences not listed in the "Guide."

ANNUAL FUEL COSTS

Fuel costs are continually changing and vary considerably by area. Use the chart (FIG. 1-4 in Chapter 1) to estimate annual costs using fuel prices in your area. These costs are based on 15,000 miles of driving per year. The annual fuel cost displayed on the fuel economy label of 1991 cars is based on 15,000 miles of driving and costs of $1.30 per gallon for premium unleaded gasoline, $1.10 per gallon for regular unleaded gasoline, or $1.05 per gallon for diesel fuel. It is suggested that you use the chart to figure your costs by first calculating the cost for the estimated highway mpg for your car, and then the cost for the estimated city mpg. Multiply each value by the percent of your total driving that is done in the city and the percent that is done on the highway, and add the two values together to yield the total cost. For example, for someone who did 80 percent city driving and 20 percent highway driving and who was considering purchasing a car which was listed as achieving 35 mpg for highway driving and 21 mpg for city driving, the calculation would appear as:

City driving mpg . . . . $786 and
highway driving mpg . . . . $471 assuming
gasoline cost $1.10 per gallon.
Then, ($786 × 0.8) + ($471 × 0.2) = $722,
the total annual cost for fuel.

The reader is cautioned that simply averaging the mpg for city and highway driving and then looking up a single value in the table will yield an incorrect answer.

Because of the constant changes in the price of gasoline, fuel economy labels may use different values in estimating the annual fuel cost.

TWO SEATERS

ACURA	TRANS	CITY	HWY	ENG/CYL	
NSX	L4	18	24	3.0/6	P
	M5	18	24	3.0/6	P

ALFA ROMEO

SPIDER	A3	22	25	2.0/4	P
	M5	22	30	2.0/4	P

BUICK

REATTA	L4	18	27	3.8/6	

CADILLAC

ALLANTE	L4	15	22	4.5/8	GP

CHEVROLET

CORVETTE	L4	16	24	5.7/8	P
	M6	16	26	5.7/8	P
	M6	16	25	5.7/8	P ★

CHRYSLER

TC BY MASERATI CONVERTIBLE	L4	18	24	3.0/6	P

FERRARI

F40	M5	12	17	2.9/8	TGP
TESTAROSSA	M5	10	15	4.9/12	G

GEO

METRO LSI CONVERTIBLE	A3	32	37	1.0/3	
	M5	41	46	1.0/3	

HONDA

CIVIC CRX	L4	29	34	1.5/4	
	M5	32	36	1.5/4	
	M5	28	33	1.6/4	
CIVIC CRX HF	M5	43	49	1.5/4	
	M5	49	52	1.5/4 ★	

JAGUAR

JAGUAR XJ-S CONVERTIBLE	A3	13	17	5.3/12	GP

LAMBORGHINI

DB132/DIABLO	M5	9	14	5.7/12	GP

LOTUS

LOTUS ESPRIT TURBO	M5	17	27	2.2/4	TP

MAZDA

MAZDA RX-7	L4	17	23	1.3/2	R
	M5	17	25	1.3/2	R
	M5	16	24	1.3/2	TR
MX-5 MIATA	L4	24	28	1.6/4	
	M5	25	30	1.6/4	

MERCEDES-BENZ

300SL	A5	16	23	3.0/6	GP
	M5	16	22	3.0/6	GP
500SL	A4	14	18	5.0/8	GP

NISSAN MOTOR COMPANY, LTD.

300ZX	L4	19	24	3.0/6	P
	L4	18	24	3.0/6	TP
	M5	18	24	3.0/6	P
	M5	18	24	3.0/6	TP

TWO SEATERS (cont'd)

	TRANS	CITY	HWY	ENG/CYL	
TOYOTA					
MR2	M5	20	27	2.0/4	TP
	L4	22	29	2.2/4	
	M5	22	28	2.2/4	

MINICOMPACT CARS

LINCOLN-MERCURY	TRANS	CITY	HWY	ENG/CYL	
CAPRI	L4	23	27	1.6/4	
	M5	25	31	1.6/4	
	M5	23	28	1.6/4	T

NISSAN MOTOR COMPANY, LTD.

NX COUPE	L4	27	36	1.6/4	
	M5	28	38	1.6/4	
	L4	23	29	2.0/4	
	M5	23	31	2.0/4	

	TRANS	CITY	HWY	ENG/CYL	
240SX	L4	21	27	2.4/4	P
	M5	22	28	2.4/4	P

PORSCHE

	TRANS	CITY	HWY	ENG/CYL	
911 CARRERA 4/2	M5	16	24	3.6/6	GP
	M5	15	22	3.6/6	GP *
	S4	16	22	3.6/6	GP
928 S4	A4	15	19	5 0/8	GP
	M5	13	19	5.0/8	GP
944 S2	M5	17	26	3.0/4	P

TOYOTA

	TRANS	CITY	HWY	ENG/CYL
CELICA CONVERTIBLE	L4	23	30	2.2/4
	M5	22	30	2.2/4

VOLKSWAGEN

	TRANS	CITY	HWY	ENG/CYL
CABRIOLET	A3	23	28	1.8/4
	M5	25	32	1.8/4

SUBCOMPACT CARS

ACURA

	TRANS	CITY	HWY	ENG/CYL
INTEGRA	L4	23	28	1.8/4
	M5	24	28	1.8/4

AUDI

	TRANS	CITY	HWY	ENG/CYL
COUPE QUATTRO	M5	17	24	2.3/5
80 QUATTRO	M5	18	24	2.3/5
80/90	L4	18	25	2.3/5
	M5	20	26	2.3/5
90 QUATTRO 20V	M5	17	24	2.3/5

SUBCOMPACT CARS (cont'd)

	TRANS	CITY	HWY	ENG/CYL	
BMW					
M3	M5	17	29	2.3/4	P

	TRANS	CITY	HWY	ENG/CYL	
318IS	M5	22	27	1.8/4	P
325I CONVERTIBLE	L4	18	23	2.5/6	
	M5	17	24	2.5/6	
325I/325IS	L4	18	22	2.5/6	
	M5	18	23	2.5/6	
325IX	L4	19	22	2.5/6	
	M5	17	23	2.5/6	
850I	L4	12	18	5.0/12	G
	M6	12	19	5.0/12	G

CHEVROLET

	TRANS	CITY	HWY	ENG/CYL	
CAMARO	L4	18	27	3.1/6	
	M5	17	27	3.1/6	
	L4	17	24	5.0/8	P
	L4	17	26	5.0/8	
	M5	16	26	5.0/8	P
	M5	17	26	5.0/8	
	L4	16	24	5.7/8	P
CAVALIER	L3	23	32	2.2/4	
	M5	24	35	2.2/4	
	L3	20	27	3.1/6	
	M5	19	28	3.1/6	

CHRYSLER

	TRANS	CITY	HWY	ENG/CYL	
LEBARON CONVERTIBLE	L3	21	26	2.5/4	
	L3	19	24	2.5/4	TP
	M5	20	27	2.5/4	TP
	L4	20	26	3.0/6	
	M5	18	27	3.0/6	

DAIHATSU MOTOR CO., LTD.

	TRANS	CITY	HWY	ENG/CYL
CHARADE	M5	38	42	1.0/3
	A3	30	32	1.3/4
	M5	35	38	1.3/4

DODGE

	TRANS	CITY	HWY	ENG/CYL	
COLT	L3	28	31	1.5/4	
	M4	31	36	1.5/4	
	M5	29	35	1.5/4	
DAYTONA	L3	23	28	2.5/4	
	L3	19	24	2.5/4	TP
	M5	24	34	2.5/4	
	M5	20	27	2.5/4	TP
	L4	20	26	3.0/6	
	M5	19	27	3.0/6	
SHADOW CONVERTIBLE	M5	20	28	2.2/4	TP
	L3	23	28	2.5/4	
	L3	19	23	2.5/4	TP

	TRANS	CITY	HWY	ENG/CYL	
	M5	24	34	2.5/4	
	M5	20	26	2.5/4	TP
STEALTH	L4	18	24	3.0/6	P
	L4	18	23	3.0/6	
	M5	19	24	3.0/6	P
	M5	18	24	3.0/6	TP
	M5	18	24	3.0/6	

EAGLE

	TRANS	CITY	HWY	ENG/CYL	
TALON	A4	19	23	2.0/4	TP
	A4	18	21	2.0/4	TP
	L4	22	27	2.0/4	
	M5	22	29	2.0/4	
	M5	21	28	2.0/4	TP
	M5	20	25	2.0/4	TP★

SUBCOMPACT CARS (cont'd)

	TRANS	CITY	HWY	ENG/CYL	
FORD					
FESTIVA	A3	31	33	1.3/4	
	M5	35	42	1.3/4	
MUSTANG	L4	21	28	2.3/4	
	M5	22	30	2.3/4	
	L4	18	25	5.0/8	
	M5	17	24	5.0/8	
GEO					
METRO	A3	36	39	1.0/3	
	M5	45	50	1.0/3	
METRO LSI	A3	36	39	1.0/3	
	M5	45	50	1.0/3	
METRO XFI	M5	53	58	1.0/3	
PRIZM	L3	26	29	1.6/4	
	L4	23	30	1.6/4	P
	M5	28	34	1.6/4	
	M5	25	30	1.6/4	P
STORM	A3	24	31	1.6/4	
	L4	23	31	1.6/4	
	M5	25	33	1.6/4	
	M5	30	36	1.6/4	★
HONDA					
CIVIC	L4	28	33	1.5/4	
	M4	33	37	1.5/4	
	M5	31	35	1.5/4	
	L4	24	29	1.6/4	
	M5	28	32	1.6/4	
PRELUDE	L4	20	26	2.0/4	
	L4	22	27	2.0/4	★

	TRANS	CITY	HWY	ENG/CYL	
	M5	22	27	2.0/4	
	M5	23	27	2.0/4	★
	L4	21	25	2.1/4	
	M5	22	26	2.1/4	
HYUNDAI MOTOR COMPANY					
EXCEL	L4	28	32	1.5/4	
	M4	29	33	1.5/4	
	M5	29	36	1.5/4	
SCOUPE	L4	25	32	1.5/4	
	M5	26	34	1.5/4	
INFINITI					
M30	L4	19	25	3.0/6	P
ISUZU					
IMPULSE	L4	24	32	1.6/4	
	M5	26	33	1.6/4	
	M5	22	28	1.6/4	TP
JAGUAR					
JAGUAR XJ-S COUPE	A3	13	18	5.3/12	GP
MERCEDES-BENZ					
190E2.3	A4	21	27	2.3/4	P
	M5	20	28	2.3/4	P
190E2.6	A4	19	24	2.6/6	P
	M5	19	27	2.6/6	P
300CE	A4	17	23	3.0/6	P
MITSUBISHI					
ECLIPSE	L4	23	30	1.8/4	
	M5	23	32	1.8/4	
	A4	19	23	2.0/4	TP
	A4	18	21	2.0/4	TP ★
	L4	22	27	2.0/4	
	M5	22	29	2.0/4	
	M5	21	28	2.0/4	TP
	M5	20	25	2.0/4	TP★

SUBCOMPACT CARS (cont'd)

	TRANS	CITY	HWY	ENG/CYL	
MIRAGE	L3	28	31	1.5/4	
	L4	26	32	1.5/4	
	M4	31	36	1.5/4	
	M5	29	35	1.5/4	
	L4	22	28	1.6/4	
	M5	23	28	1.6/4	
3000 GT	L4	18	24	3.0/6	P
	M5	19	24	3.0/6	P
	M5	18	24	3.0/6	TP

NISSAN MOTOR COMPANY, LTD.

	TRANS	CITY	HWY	ENG/CYL	
SENTRA	L3	28	33	1.6/4	
	L4	27	36	1.6/4	
	M4	29	37	1.6/4	
	M5	29	39	1.6/4	
	L4	23	30	2.0/4	
	M5	24	32	2.0/4	
300ZX 2+2	L4	19	24	3.0/6	P
	M5	18	24	3.0/6	P

PLYMOUTH

	TRANS	CITY	HWY	ENG/CYL	
COLT	L3	28	31	1.5/4	
	M4	31	36	1.5/4	
	M5	29	35	1.5/4	
LASER	L4	23	30	1.8/4	
	M5	23	32	1.8/4	
	A4	19	23	2.0/4	TP
	L4	22	27	2.0/4	
	M5	22	29	2.0/4	
	M5	21	28	2.0/4	TP

PONTIAC

	TRANS	CITY	HWY	ENG/CYL	
FIREBIRD/ TRANS AM	L4	18	27	3.1/6	
	M5	17	27	3.1/6	
	L4	17	24	5.0/8	P
	L4	17	26	5.0/8	
	M5	16	26	5.0/8	P
	M5	17	26	5.0/8	
	L4	16	24	5.7/8	P
SUNBIRD	L3	24	32	2.0/4	
	M5	26	36	2.0/4	
	L3	20	27	3.1/6	
	M5	19	28	3.1/6	
SUNBIRD CONVERTIBLE	L3	22	31	2.0/4	
	M5	24	33	2.0/4	
	L3	20	27	3.1/6	
	M5	19	28	3.1/6	

ROLLS-ROYCE MOTOR CARS LTD.

	TRANS	CITY	HWY	ENG/CYL	
BENTLEY CONTINENTAL	A3	10	13	6.8/8	GP
CORNICHE III	A3	10	13	6.8/8	GP

SAAB

	TRANS	CITY	HWY	ENG/CYL	
SAAB 900 CONVERTIBLE	A3	18	21	2.0/4	T
	M5	21	27	2.0/4	T

SATURN

	TRANS	CITY	HWY	ENG/CYL	
SC	L4	23	32	1.9/4	
	M5	24	34	1.9/4	

SUBARU

	TRANS	CITY	HWY	ENG/CYL	
JUSTY	M5	33	37	1.2/3	
	AV	33	35	1.2/3	
JUSTY 4WD	AV	31	31	1.2/3	
	M5	29	32	1.2/3	

SUBCOMPACT CARS (cont'd)

	TRANS	CITY	HWY	ENG/CYL	
XT	L4	23	29	1.8/4	
	M5	25	31	1.8/4	
	L4	20	28	2.7/6	
XT 4WD	L4	22	28	1.8/4	
	M5	24	29	1.8/4	
	L4	19	25	2.7/6	
	M5	18	25	2.7/6	

SUZUKI MOTOR CO.,LTD.

	TRANS	CITY	HWY	ENG/CYL	
SWIFT	A3	36	39	1.0/3	
	M5	45	50	1.0/3	
	A3	29	33	1.3/4	
	M5	39	43	1.3/4	
SWIFT GT	M5	28	35	1.3/4	

TOYOTA

	TRANS	CITY	HWY	ENG/CYL	
CELICA	L4	25	33	1.6/4	
	M5	28	33	1.6/4	
	M5	19	24	2.0/4	TP
	L4	22	28	2.2/4	
	L4	23	30	2.2/4	*
	M5	22	30	2.2/4	
COROLLA	L3	26	29	1.6/4	
	L4	25	33	1.6/4	
	M5	28	33	1.6/4	
	M5	25	31	1.6/4	P
SUPRA	L4	18	23	3.0/6	P
	L4	18	23	3.0/6	TP
	M5	18	23	3.0/6	P
	M5	17	23	3.0/6	TP
TERCEL	L3	26	29	1.5/4	
	M4	33	37	1.5/4	
	M5	29	35	1.5/4	

VOLKSWAGEN

	TRANS	CITY	HWY	ENG/CYL	
FOX	M4	25	32	1.8/4	
	M5	25	33	1.8/4	

COMPACT CARS

ALFA ROMEO	TRANS	CITY	HWY	ENG/CYL		
164	L4	18	25	3.0/6	P	
	M5	18	27	3.0/6	P	
	M5	17	25	3.0/6	P	★

BMW					
M5	M5	11	20	3.5/6	GP

BUICK				
SKYLARK	L3	23	33	2.3/4
	L3	22	31	2.5/4
	L3	19	26	3.3/6

CHEVROLET					
BERETTA	L3	24	31	2.2/4	
	M5	24	33	2.2/4	
	M5	23	33	2.3/4	P
	L3	20	27	3.1/6	
	M5	19	28	3.1/6	

COMPACT CARS (cont'd)

CHRYSLER	TRANS	CITY	HWY	ENG/CYL	
LEBARON	L3	23	28	2.5/4	
	L3	19	24	2.5/4	TP
	M5	20	27	2.5/4	TP
	L4	20	26	3.0/6	
	M5	19	29	3.0/6	

DODGE					
SHADOW	L3	23	27	2.2/4	
	M5	23	30	2.2/4	
	L3	23	28	2.5/4	
	L3	19	23	2.5/4	TP
	M5	24	34	2.5/4	
	M5	20	26	2.5/4	TP

EAGLE				
SUMMIT	L3	28	31	1.5/4
	L4	26	32	1.5/4
	M4	31	36	1.5/4
	M5	29	35	1.5/4

FORD					
ESCORT	L4	23	30	1.8/4	
	M5	26	31	1.8/4	
	L4	25	33	1.9/4	
	M5	29	36	1.9/4	
ESCORT FS	M5	31	41	1.9/4	
PROBE	L4	21	28	2.2/4	
	L4	19	25	2.2/4	TP
	M5	24	31	2.2/4	
	M5	21	27	2.2/4	TP
	L4	19	25	3.0/6	
	M5	19	26	3.0/6	
TEMPO	A3	22	26	2.3/4	
	M5	21	29	2.3/4	
	M5	22	32	2.3/4	★
TEMPO ALL WHEEL DRIVE	A3	20	24	2.3/4	

HONDA				
ACCORD	L4	22	28	2.2/4
	M5	24	29	2.2/4

HYUNDAI MOTOR COMPANY				
PRECIS	L4	28	32	1.5/4
	M4	29	33	1.5/4
	M5	29	36	1.5/4

INFINITI				
G20	L4	22	29	2.0/4
	M5	24	32	2.0/4

ISUZU					
STYLUS	A3	28	33	1.6/4	
	M5	25	33	1.6/4	
	M5	31	37	1.6/4	★

JAGUAR					
JAGUAR XJ6	L4	17	22	4.0/6	P

LEXUS				
ES250	L4	18	24	2.5/6
	M5	19	25	2.5/6

LINCOLN-MERCURY					
TOPAZ	A3	22	26	2.3/4	
	M5	21	29	2.3/4	
	M5	22	32	2.3/4	★
TOPAZ ALL WHEEL DRIVE	A3	20	24	2.3/4	

COMPACT CARS (cont'd)

	TRANS	CITY	HWY	ENG/CYL	
TRACER	L4	23	30	1.8/4	
	M5	26	31	1.8/4	
	L4	25	33	1.9/4	
	M5	29	36	1.9/4	

MAZDA

	TRANS	CITY	HWY	ENG/CYL	
323 PROTEGE 4X4	L4	21	26	1.8/4	
	M5	24	29	1.8/4	
323/323 PROTEGE	L4	26	33	1.6/4	
	M5	29	37	1.6/4	
	L4	24	29	1.8/4	
	L4	24	31	1.8/4	★
	M5	25	30	1.8/4	
	M5	28	36	1.8/4	★

MERCEDES-BENZ

	TRANS	CITY	HWY	ENG/CYL	
300D 2.5 TURBO	A4	27	33	2.5/5	TD
300E	A4	18	23	3.0/6	P
300E 2.6	A4	19	24	2.6/6	P
300E-4MATIC	A4	17	21	3.0/6	GP
300SE	A4	16	20	3.0/6	GP
350SD TURBO...	A4	22	25	3.5/6	TD
560SEC	A4	14	17	5.6/8	GP

MITSUBISHI

	TRANS	CITY	HWY	ENG/CYL	
GALANT	L4	21	26	2.0/4	
	L4	19	23	2.0/4	
	L4	21	27	2.0/4	★
	M5	21	27	2.0/4	
	M5	20	23	2.0/4	★
	M5	19	25	2.0/4	TP
	M5	22	29	2.0/4	

NISSAN MOTOR COMPANY, LTD.

	TRANS	CITY	HWY	ENG/CYL	
STANZA	L4	21	27	2.4/4	
	M5	22	29	2.4/4	

OLDSMOBILE

	TRANS	CITY	HWY	ENG/CYL	
CUTLASS CALAIS	L3	23	33	2.3/4	
	M5	23	33	2.3/4	P
	L3	22	31	2.5/4	
	M5	22	34	2.5/4	
	L3	19	26	3.3/6	

PEUGEOT

	TRANS	CITY	HWY	ENG/CYL	
405 SEDAN........	L4	21	25	1.9/4	
	M5	21	28	1.9/4	
	M5	20	28	1.9/4	P

PLYMOUTH

	TRANS	CITY	HWY	ENG/CYL	
SUNDANCE.........	L3	23	27	2.2/4	
	M5	23	30	2.2/4	
	L3	23	28	2.5/4	
	L3	19	23	2.5/4	TP
	M5	24	34	2.5/4	
	M5	20	26	2.5/4	TP

PONTIAC

	TRANS	CITY	HWY	ENG/CYL	
GRAND AM.........	L3	23	33	2.3/4	
	M5	23	33	2.3/4	P
	L3	22	31	2.5/4	
	M5	22	33	2.5/4	
LEMANS	L3	27	33	1.6/4	
	M4	29	38	1.6/4	
	M5	31	40	1.6/4	
	L3	23	30	2.0/4	
	M5	25	31	2.0/4	

COMPACT CARS (cont'd)

	TRANS	CITY	HWY	ENG/CYL	

ROVER GROUP

	TRANS	CITY	HWY	ENG/CYL	
STERLING 827...	L4	18	22	2.7/6	
	M5	19	23	2.7/6	

SAAB

	TRANS	CITY	HWY	ENG/CYL	
SAAB 900	A3	19	24	2.0/4	T
	M5	22	29	2.0/4	T
	M5	20	28	2.0/4	TP
	A3	19	22	2.1/4	
	M5	20	26	2.1/4	

SATURN

	TRANS	CITY	HWY	ENG/CYL	
SL.......................	L4	23	32	1.9/4	
	L4	26	35	1.9/4	★
	M5	24	34	1.9/4	
	M5	27	37	1.9/4	★

SUBARU

	TRANS	CITY	HWY	ENG/CYL	
LEGACY	L4	21	28	2.2/4	
	M5	22	29	2.2/4	
LEGACY TURBO	L4	19	25	2.2/4	T
LEGACY 4WD	L4	20	26	2.2/4	
	M5	20	27	2.2/4	

	TRANS	CITY	HWY	ENG/CYL	
LEGACY 4WD					
TURBO	L4	18	24	2.2/4	T
	M5	19	25	2.2/4	T
LOYALE	L3	24	26	1.8/4	
	M5	25	32	1.8/4	

TOYOTA

	TRANS	CITY	HWY	ENG/CYL	
CAMRY	L4	24	30	2.0/4	
	L4	21	24	2.0/4	*
	L4	24	31	2.0/4	
	M5	26	34	2.0/4	
	L4	18	24	2.5/6	
	M5	19	25	2.5/6	
CRESSIDA	L4	19	24	3.0/6	P

VOLKSWAGEN

	TRANS	CITY	HWY	ENG/CYL	
GOLF/GTI	A3	23	28	1.8/4	
	M5	25	32	1.8/4	
GTI 16V	M5	22	28	2.0/4	
JETTA	M5	37	43	1.6/4	D
	A3	23	28	1.8/4	
	M5	25	32	1.8/4	
JETTA GLI 16V	M5	22	28	2.0/4	

VOLVO

	TRANS	CITY	HWY	ENG/CYL	
COUPE	A4	19	22	2.3/4	T
240	A4	20	25	2.3/4	
	M5	21	28	2.3/4	

MID-SIZE CARS

AUDI	TRANS	CITY	HWY	ENG/CYL	
V8	L4	15	20	3.6/8	G
	M5	14	20	3.6/8	G
100	L4	18	23	2.3/5	
100 QUATTRO	M5	18	24	2.3/5	

MID-SIZE CARS (cont'd)

	TRANS	CITY	HWY	ENG/CYL	
200	A3	18	22	2.2/5	T

200 QUATTRO	TRANS	CITY	HWY	ENG/CYL	
20V	M5	18	24	2.2/5	T

BMW

	TRANS	CITY	HWY	ENG/CYL	
750IL	L4	12	18	5.0/12	G

BUICK

	TRANS	CITY	HWY	ENG/CYL	
CENTURY	L3	22	31	2.5/4	
	L3	19	26	3.3/6	
	L4	19	30	3.3/6	
REGAL	L4	19	30	3.1/6	
	L4	19	28	3.8/6	
RIVIERA	L4	18	27	3.8/6	

CADILLAC

	TRANS	CITY	HWY	ENG/CYL	
ELDORADO	L4	16	26	4.9/8	P
SEVILLE	L4	16	26	4.9/8	P

CHEVROLET

	TRANS	CITY	HWY	ENG/CYL	
CORSICA	L3	24	31	2.2/4	
	M5	24	33	2.2/4	
	L3	20	27	3.1/6	
	M5	19	28	3.1/6	
LUMINA	L3	21	28	2.5/4	
	L3	19	27	3.1/6	
	L4	19	30	3.1/6	

CHRYSLER

	TRANS	CITY	HWY	ENG/CYL	
NEW YORKER	L4	19	26	3.3/6	

CX AUTOMOTIVE INC

	TRANS	CITY	HWY	ENG/CYL	
XM-V6	M5	16	22	3.0/6	GP

DODGE

	TRANS	CITY	HWY	ENG/CYL	
DYNASTY	L3	21	26	2.5/4	
	L4	20	26	3.0/6	
	L4	19	26	3.3/6	
SPIRIT	M5	19	27	2.2/4	TP
	L3	23	27	2.5/4	
	L3	19	24	2.5/4	TP
	M5	24	34	2.5/4	
	M5	20	27	2.5/4	TP
	L4	20	26	3.0/6	

FORD

	TRANS	CITY	HWY	ENG/CYL	
TAURUS	L4	20	29	3.0/6	
	M5	18	26	3.0/6	P
THUNDERBIRD	L4	19	27	3.8/6	
	L4	17	23	3.8/6	P
	M5	17	24	3.8/6	P
	L4	18	24	5.0/8	

HYUNDAI MOTOR COMPANY

	TRANS	CITY	HWY	ENG/CYL	
SONATA	L4	21	26	2.4/4	
	M5	21	28	2.4/4	
	A4	18	24	3.0/6	

INFINITI

Q45	L4	16	22	4.5/8	GP
Q45 FULL-ACTIVE SUSPENSION	L4	14	19	4.5/8	GP

LEXUS

LS400	L4	18	23	4.0/8	P

LINCOLN-MERCURY

COUGAR	L4	19	27	3.8/6	
	L4	18	24	5.0/8	

MID-SIZE CARS (cont'd)

	TRANS	CITY	HWY	ENG/CYL	
MARK VII	L4	17	24	5.0/8	
SABLE	L4	20	29	3.0/6	

MAZDA

626/MX-6	L4	22	28	2.2/4	
	L4	19	25	2.2/4	TP
	M5	24	31	2.2/4	
	M5	21	28	2.2/4	TP
929	L4	18	22	3.0/6	
	L4	19	23	3.0/6	★

MERCEDES-BENZ

300SEL	A4	16	20	3.0/6	GP
350SDL TURBO	A4	22	25	3.5/6	TD
420SEL	A4	15	19	4.2/8	GP
560SEL	A4	14	17	5.6/8	GP

NISSAN MOTOR COMPANY, LTD.

MAXIMA	L4	19	26	3.0/6	P
	M5	20	26	3.0/6	P

OLDSMOBILE

CUTLASS CIERA	L3	22	31	2.5/4	
	L3	19	26	3.3/6	
	L4	19	29	3.3/6	
CUTLASS SUPREME	L3	21	29	2.3/4	
	L4	19	30	3.1/6	

TROFEO/TORONADO	L4	18	27	3.8/6	

PLYMOUTH

ACCLAIM	L3	23	28	2.5/4	
	M5	24	34	2.5/4	
	L4	20	26	3.0/6	

PONTIAC

GRAND PRIX	L3	21	29	2.3/4	
	L4	19	30	3.1/6	
6000	L3	22	31	2.5/4	
	L3	20	27	3.1/6	
	L4	19	30	3.1/6	

ROLLS-ROYCE MOTOR CARS LTD.

BENTLEY EIGHT/MULSANNE S	A3	10	13	6.8/8	GP
BENTLEY TURBO R	A3	10	13	6.8/8	TGP
SILVER SPIRIT II/SILVER SPUR	A3	10	13	6.8/8	GP

VOLKSWAGEN

PASSAT	L4	20	29	2.0/4	
	M5	21	30	2.0/4	

VOLVO

740	A4	19	22	2.3/4	T
	L4	20	26	2.3/4	
	M5	22	28	2.3/4	
	M5	20	26	2.3/4	T
940 GLE 16-VALVE	L4	18	23	2.3/4	
940 SE	A4	19	22	2.3/4	T
940 TURBO	A4	19	22	2.3/4	T

LARGE CARS

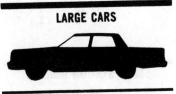

BUICK	TRANS	CITY	HWY	ENG/CYL	
LESABRE	L4	19	28	3.8/6	
PARK AVENUE	L4	18	27	3.8/6	

CADILLAC

	TRANS	CITY	HWY	ENG/CYL	
BROUGHAM........	L4	16	25	5.0/8	
	L4	15	22	5.7/8	G
FLEETWOOD/ DEVILLE.............	L4	16	26	4.9/8	P

CHEVROLET

CAPRICE	L4	17	26	5.0/8	

CHRYSLER

NEW YORKER FIFTH AVENUE/ IMPE..................	L4	18	26	3.3/6	
	L4	17	25	3.8/6	

DODGE

MONACO	L4	18	27	3.0/6	

EAGLE

PREMIER............	L4	18	27	3.0/6	

FORD

LTD CROWN VICTORIA	L4	17	24	5.0/8	

LINCOLN-MERCURY

GRAND MARQUIS	L4	17	24	5.0/8	
TOWN CAR........	L4	17	23	4.6/8	

OLDSMOBILE

EIGHTY-EIGHT....	L4	19	28	3.8/6	
NINETY-EIGHT/ TOURING	L4	18	27	3.8/6	

PONTIAC

BONNEVILLE......	L4	19	28	3.8/6	

SAAB

SAAB 9000........	L4	17	25	2.3/4	
	L4	17	24	2.3/4	T
	M5	20	26	2.3/4	
	M5	19	25	2.3/4	T

SMALL STATION WAGONS

CHEVROLET

	TRANS	CITY	HWY	ENG/CYL
CAVALIER WAGON..............	L3	24	31	2.2/4
	M5	24	33	2.2/4
	L3	20	27	3.1/6

DODGE

COLT VISTA.......	L3	22	23	2.0/4
	M5	22	29	2.0/4

SMALL STATION WAGONS (cont'd)

	TRANS	CITY	HWY	ENG/CYL

FORD

ESCORT WAGON	L4	25	33	1.9/4
	M5	29	36	1.9/4

HONDA

CIVIC WAGON....	L4	28	32	1.5/4
	M5	31	35	1.5/4
CIVIC WAGON 4WD..................	L4	25	27	1.6/4
	M5	24	26	1.6/4

LINCOLN-MERCURY

TRACER WAGON	L4	25	33	1.9/4
	M5	29	36	1.9/4

MITSUBISHI

SPACE WAGON ..	L3	22	23	2.0/4

PLYMOUTH

COLT VISTA.......	L3	22	23	2.0/4
	M5	22	29	2.0/4

SUBARU

LEGACY WAGON	L4	21	27	2.2/4
	M5	22	29	2.2/4
LEGACY WAGON 4WD..................	L4	20	26	2.2/4
	M5	20	26	2.2/4
LOYALE WAGON	L3	24	26	1.8/4
	M5	25	30	1.8/4

TOYOTA

CAMRY WAGON .	L4	24	30	2.0/4
	L4	18	24	2.5/6
COROLLA ALL- TRAC WAGON	L4	23	29	1.6/4
	M5	22	27	1.6/4

COROLLA
WAGON...............

	TRANS	CITY	HWY	ENG/CYL	
	L3	26	29	1.6/4	
	M5	28	33	1.6/4	

MID-SIZE STATION WAGONS

AUDI	TRANS	CITY	HWY	ENG/CYL	
200 QUATTRO 20V WAGON	M5	18	24	2.2/5	T

BUICK

CENTURY WAGON...............	L3	21	28	2.5/4	
	L3	19	25	3.3/6	
	L4	19	29	3.3/6	

FORD

TAURUS WAGON...............	L4	20	29	3.0/6	

LINCOLN-MERCURY

SABLE WAGON..	L4	20	29	3.0/6	

MERCEDES-BENZ

300TE.................	A4	17	22	3.0/6	GP

MID-SIZE STATION WAGONS (cont'd)

	TRANS	CITY	HWY	ENG/CYL	
300TE-4MATIC..	A4	16	20	3.0/6	GP

OLDSMOBILE

CUTLASS CRUISER.............	L3	21	28	2.5/4	
	L3	19	25	3.3/6	
	L4	19	29	3.3/6	

PEUGEOT

405 SPORTS WAGON...............	L4	20	25	1.9/4	
	M5	20	27	1.9/4	
505 STATION WAGON...............	L4	19	21	2.2/4	
	L4	18	21	2.2/4	TP

	M5	18	22	2.2/4	

PONTIAC

6000 WAGON	L3	21	28	2.5/4	
	L4	19	30	3.1/6	

VOLKSWAGEN

PASSAT WAGON	L4	20	29	2.0/4	
	M5	21	30	2.0/4	

VOLVO

240 WAGON.......	A4	20	25	2.3/4	
	M5	22	28	2.3/4	
740 WAGON.......	A4	19	22	2.3/4	T
	L4	20	26	2.3/4	
	M5	22	28	2.3/4	
	M5	20	26	2.3/4	T
940 GLE 16-VALVE WAGON..	L4	18	23	2.3/4	
940 SE WAGON.	A4	19	22	2.3/4	T
940 TURBO WAGON...............	A4	19	22	2.3/4	T

LARGE STATION WAGONS

BUICK	TRANS	CITY	HWY	ENG/CYL	
ROADMASTER WAGON...............	L4	16	25	5.0/8	

CHEVROLET

CAPRICE WAGON...............	L4	16	25	5.0/8	

FORD

LTD CROWN VICTORIA WAGON...............	L4	17	24	5.0/8	

LINCOLN-MERCURY

GRAND MARQUIS WAGON...............	L4	17	24	5.0/8	

OLDSMOBILE

CUSTOM CRUISER.............	L4	16	25	5.0/8	

2WD SMALL PICKUPS

M5 19 23 3.0/6

CHEVROLET	TRANS	CITY	HWY	ENG/CYL
S10 PICKUP 2WD	L4	21	28	2.5/4
	M5	23	27	2.5/4
	M5	19	25	2.8/6
	L4	18	24	4.3/6
	M5	17	23	4.3/6

DODGE	TRANS	CITY	HWY	ENG/CYL
RAM 50 2WD	L4	19	23	2.4/4
	M5	19	24	2.4/4

FORD	TRANS	CITY	HWY	ENG/CYL
RANGER PICKUP 2WD	L4	21	24	2.3/4
	M5	24	29	2.3/4
	L4	18	23	3.0/6
	M5	19	25	3.0/6
	L4	17	23	4.0/6
	M5	18	23	4.0/6

GMC	TRANS	CITY	HWY	ENG/CYL
SONOMA PICKUP 2WD	L4	21	28	2.5/4
	M5	23	27	2.5/4
	M5	19	25	2.8/6
	L4	18	24	4.3/6
	M5	17	23	4.3/6

ISUZU	TRANS	CITY	HWY	ENG/CYL
PICKUP 2WD	M5	22	24	2.3/4
	L4	19	23	2.6/4
	M5	19	24	2.6/4

MITSUBISHI	TRANS	CITY	HWY	ENG/CYL
TRUCK 2WD	L4	19	23	2.4/4
	M5	19	24	2.4/4

NISSAN MOTOR COMPANY, LTD.	TRANS	CITY	HWY	ENG/CYL
TRUCK 2WD	L4	21	26	2.4/4
	M5	23	27	2.4/4
	L4	18	24	3.0/6

4WD SMALL PICKUPS

FORD	TRANS	CITY	HWY	ENG/CYL
RANGER PICKUP 4WD	M5	22	25	2.3/4
	L4	17	21	2.9/6
	M5	18	22	2.9/6
	L4	16	20	4.0/6
	M5	17	21	4.0/6

2WD LARGE PICKUPS

CHEVROLET	TRANS	CITY	HWY	ENG/CYL	
C1500 PICKUP 2WD	L4	17	22	4.3/6	
	M4C	18	20	4.3/6	
	M5	17	23	4.3/6	
	L4	15	19	5.0/8	
	M5	15	20	5.0/8	
	L4	15	19	5.7/8	
	M4C	13	16	5.7/8	
	M5	14	19	5.7/8	
	M5C	13	18	5.7/8	
	L4	18	24	6.2/8	D
	M4C	19	21	6.2/8	D
	L4	10	12	7.4/8	
C2500 PICKUP 2WD	L4	16	21	4.3/6	
	M4C	17	20	4.3/6	
	M5	17	23	4.3/6	
	L4	14	18	5.0/8	

TRANS	CITY	HWY	ENG/CYL	
M5	15	20	5.0/8	
L4	14	19	5.7/8	
M4C	13	15	5.7/8	
M5	12	17	5.7/8	
M5C	13	18	5.7/8	
L4	17	22	6.2/8	D
M4C	19	20	6.2/8	D

DODGE

DAKOTA
PICKUP 2WD

TRANS	CITY	HWY	ENG/CYL
M5	21	27	2.5/4
L4	16	21	3.9/6
M5	16	22	3.9/6
L4	14	19	5.2/8

D100/D150
PICKUP 2WD

TRANS	CITY	HWY	ENG/CYL
L3	15	17	3.9/6
L4	16	21	3.9/6
M4C	14	17	3.9/6
M5	13	18	3.9/6
A4	13	16	5.2/8
L3	13	15	5.2/8
L4	14	18	5.2/8
M4C	13	15	5.2/8
A4	11	13	5.9/8
M4C	10	13	5.9/8

D250 PICKUP
2WD

TRANS	CITY	HWY	ENG/CYL
L3	15	16	3.9/6
M4C	14	17	3.9/6
A4	13	15	5.2/8
L4	13	17	5.2/8
M4C	12	14	5.2/8
A4	11	13	5.9/8
M4C	10	13	5.9/8

FORD

F150 PICKUP
2WD

TRANS	CITY	HWY	ENG/CYL
L4	15	20	4.9/6
M4C	16	18	4.9/6
M5	16	20	4.9/6
L4	14	18	5.0/8
M4C	14	15	5.0/8
M5	14	18	5.0/8
L4	12	17	5.8/8

2WD LARGE PICKUPS (cont'd)

F250 PICKUP
2WD

TRANS	CITY	HWY	ENG/CYL
L4	15	20	4.9/6
M4C	16	18	4.9/6
M5	16	20	4.9/6
L4	14	18	5.0/8
M4C	14	15	5.0/8
M5	14	17	5.0/8
L4	12	16	5.8/8

GMC

C1500 SIERRA
2WD

TRANS	CITY	HWY	ENG/CYL	
L4	17	22	4.3/6	
M4C	18	20	4.3/6	
M5	17	23	4.3/6	
L4	15	19	5.0/8	
M5	15	20	5.0/8	
L4	15	19	5.7/8	
M4C	13	16	5.7/8	
M5	14	19	5.7/8	
M5C	13	18	5.7/8	
L4	18	24	6.2/8	D
M4C	19	21	6.2/8	D

C2500 SIERRA
2WD

TRANS	CITY	HWY	ENG/CYL	
L4	16	21	4.3/6	
M4C	17	20	4.3/6	
M5	17	23	4.3/6	
L4	14	18	5.0/8	
M5	15	20	5.0/8	
L4	14	19	5.7/8	
M4C	13	15	5.7/8	
M5	12	17	5.7/8	
M5C	13	18	5.7/8	
L4	17	22	6.2/8	D
M4C	19	20	6.2/8	D

ISUZU

PICKUP 2WD
1TON

TRANS	CITY	HWY	ENG/CYL
M5	17	21	2.6/4

JEEP

COMANCHE
PICKUP 2WD

TRANS	CITY	HWY	ENG/CYL
M4	20	23	2.5/4
M5	19	23	2.5/4
L4	15	21	4.0/6
M5	17	22	4.0/6

MAZDA

B2200/B2600I

TRANS	CITY	HWY	ENG/CYL
L4	20	25	2.2/4
M4	21	25	2.2/4

	TRANS	CITY	HWY	ENG/CYL	
	M5	21	26	2.2/4	
	L4	21	26	2.2/4	*
	M5	24	26	2.2/4	*
	L4	19	24	2.6/4	
	M5	19	23	2.6/4	

TOYOTA

	TRANS	CITY	HWY	ENG/CYL
TRUCK 2WD	A4	22	23	2.4/4
	M5	22	26	2.4/4
	L4	18	23	3.0/6
	M5	18	23	3.0/6
1-TON TRUCK 2WD	L4	18	23	3.0/6
	M5	18	23	3.0/6

4WD LARGE PICKUPS

CHEVROLET	TRANS	CITY	HWY	ENG/CYL	
K1500 PICKUP 4WD	L4	16	19	4.3/6	
	M4C	15	17	4.3/6	
	M5	16	20	4.3/6	
	L4	14	17	5.0/8	
	M5	14	18	5.0/8	
	L4	13	16	5.7/8	
	M4C	12	14	5.7/8	
	M5	13	17	5.7/8	
	L4	16	20	6.2/8	D
	M4C	17	18	6.2/8	D
K2500 PICKUP 4WD	L4	15	19	4.3/6	
	M4C	15	17	4.3/6	
	M5	16	20	4.3/6	
	L4	14	17	5.0/8	
	M5	14	18	5.0/8	
	L4	13	16	5.7/8	
	M4C	12	14	5.7/8	
	M5	12	16	5.7/8	
	L4	16	21	6.2/8	D
	M4C	17	18	6.2/8	D

S10 PICKUP	TRANS	CITY	HWY	ENG/CYL
4WD	L4	17	22	4.3/6
	M5	16	20	4.3/6

DODGE

DAKOTA PICKUP 4WD	TRANS	CITY	HWY	ENG/CYL
	L4	15	19	3.9/6
	M5	15	20	3.9/6
	L4	13	17	5.2/8
POWER RAM 50 4WD	L4	18	20	2.4/4
	M5	19	22	2.4/4
	L4	17	20	3.0/6
	M5	17	22	3.0/6
W100/W150 PICKUP 4WD	A4	14	15	3.9/6
	M4C	13	16	3.9/6
	A4	12	14	5.2/8
	M4C	11	14	5.2/8
	A4	11	13	5.9/8
	M4C	9	12	5.9/8
W250 PICKUP 4WD	A4	12	14	5.2/8
	M4C	11	14	5.2/8
	A4	10	13	5.9/8
	M4C	9	13	5.9/8

FORD

F150 PICKUP 4WD	TRANS	CITY	HWY	ENG/CYL
	L4	15	19	4.9/6
	M4C	16	18	4.9/6
	M5	15	19	4.9/6
	L4	13	17	5.0/8
	M4C	13	15	5.0/8
	M5	13	16	5.0/8
	L4	11	16	5.8/8
F250 PICKUP 4WD	M5C	14	18	4.9/6
	L4	13	17	5.0/8
	M4C	12	13	5.0/8
	M5	13	16	5.0/8
	L4	11	16	5.8/8

4WD LARGE PICKUPS (cont'd)

	TRANS	CITY	HWY	ENG/CYL

GMC

K1500 SIERRA 4WD	TRANS	CITY	HWY	ENG/CYL
	L4	16	19	4.3/6

<table>
<tr><td>M4C</td><td>15</td><td>17</td><td>4.3/6</td><td></td></tr>
</table>

	TRANS	CITY	HWY	ENG/CYL	
	M4C	15	17	4.3/6	
	M5	16	20	4.3/6	
	L4	14	17	5.0/8	
	M5	14	18	5.0/8	
	L4	13	16	5.7/8	
	M4C	12	14	5.7/8	
	M5	13	17	5.7/8	
	L4	16	20	6.2/8	D
	M4C	17	18	6.2/8	D
K2500 SIERRA 4WD	L4	15	19	4.3/6	
	M4C	15	17	4.3/6	
	M5	16	20	4.3/6	
	L4	14	17	5.0/8	
	M5	14	18	5.0/8	
	L4	13	16	5.7/8	
	M4C	12	14	5.7/8	
	M5	12	16	5.7/8	
	L4	16	21	6.2/8	D
	M4C	17	18	6.2/8	D
SONOMA PICKUP 4WD	L4	17	22	4.3/6	
	M5	16	20	4.3/6	

ISUZU

	TRANS	CITY	HWY	ENG/CYL
PICKUP 4WD	M5	16	20	2.6/4

JEEP

	TRANS	CITY	HWY	ENG/CYL
COMANCHE PICKUP 4WD	M4	18	22	2.5/4
	M5	18	22	2.5/4
	L4	14	19	4.0/6
	M5	17	21	4.0/6

MAZDA

	TRANS	CITY	HWY	ENG/CYL
B2600I 4X4	L4	17	20	2.6/4
	M5	18	20	2.6/4

MITSUBISHI

	TRANS	CITY	HWY	ENG/CYL
TRUCK 4WD	L4	18	20	2.4/4
	M5	19	22	2.4/4
	M5	17	22	3.0/6

NISSAN MOTOR COMPANY, LTD.

	TRANS	CITY	HWY	ENG/CYL
TRUCK 4WD	M5	19	22	2.4/4
	L4	16	19	3.0/6
	M5	15	19	3.0/6

TOYOTA

	TRANS	CITY	HWY	ENG/CYL
TRUCK 4WD	L4	18	19	2.4/4
	M5	19	22	2.4/4
	L4	15	19	3.0/6
	M5	16	19	3.0/6

CARGO VANS

CHEVROLET	TRANS	CITY	HWY	ENG/CYL
ASTRO AWD (CARGO)	L4	17	22	4.3/6

CARGO VANS (cont'd)

	TRANS	CITY	HWY	ENG/CYL	
ASTRO 2WD (CARGO)	L4	17	22	4.3/6	
G10/20 VAN 2WD	L4	16	22	4.3/6	
	L4	14	18	5.0/8	
	L4	15	19	5.7/8	
	L4	17	22	6.2/8	D
G30 VAN 2WD	L4	16	21	4.3/6	

DODGE

	TRANS	CITY	HWY	ENG/CYL
B150/B250 VAN 2WD	A4	15	18	3.9/6
	L3	15	17	3.9/6
	M5	13	18	3.9/6
	A4	13	16	5.2/8
	L3	13	16	5.2/8
	L4	13	17	5.2/8
	A4	11	13	5.9/8
B350 VAN 2WD	L3	15	16	3.9/6
	A4	13	15	5.2/8
	L4	13	17	5.2/8
	A4	11	13	5.9/8

FORD

	TRANS	CITY	HWY	ENG/CYL
AEROSTAR VAN	L4	18	23	3.0/6
	M5	19	24	3.0/6
	L4	16	22	4.0/6
AEROSTAR VAN ALL WHEEL DRIVE	L4	16	20	4.0/6
E150 ECONOLINE 2WD	A3	14	16	4.9/6

	TRANS	CITY	HWY	ENG/CYL	
	L4	15	20	4.9/6	
	L4	13	17	5.0/8	
	L4	11	16	5.8/8	

E250 ECONOLINE 2WD

	TRANS	CITY	HWY	ENG/CYL
	A3	14	15	4.9/6
	L4	14	17	4.9/6
	L4	13	17	5.0/8
	L4	11	15	5.8/8

GMC

G15/25 VANDURA 2WD

	TRANS	CITY	HWY	ENG/CYL	
	L4	16	22	4.3/6	
	L4	14	18	5.0/8	
	L4	15	19	5.7/8	
	L4	17	22	6.2/8	D

G35 VANDURA 2WD

	L4	16	21	4.3/6

SAFARI AWD (CARGO)

	L4	17	22	4.3/6

SAFARI 2WD (CARGO)

	L4	17	22	4.3/6

PASSENGER VANS

CHEVROLET	TRANS	CITY	HWY	ENG/CYL
ASTRO AWD (PASSENGER)	L4	16	19	4.3/6

PASSENGER VANS (cont'd)

	TRANS	CITY	HWY	ENG/CYL	
ASTRO 2WD (PASSENGER)	L4	16	21	4.3/6	
G10/20 SPORTVAN 2WD	L4	15	19	4.3/6	
	L4	14	18	5.0/8	
	L4	14	19	5.7/8	
	L4	16	22	6.2/8	D

G30 SPORTVAN 2WD

	L4	13	17	5.7/8

DODGE

B150/B250 WAGON 2WD

	TRANS	CITY	HWY	ENG/CYL
	A4	15	17	3.9/6
	L3	15	16	3.9/6
	M5	12	16	3.9/6
	A4	12	15	5.2/8
	L3	13	16	5.2/8
	L4	13	17	5.2/8
	A4	11	13	5.9/8

B350 WAGON 2WD

	A4	12	15	5.2/8
	L4	13	17	5.2/8
	A4	10	12	5.9/8

FORD

AEROSTAR WAGON

	L4	17	23	3.0/6
	M5	18	23	3.0/6
	L4	16	21	4.0/6

AEROSTAR WAGON ALL WHEEL DRI

	L4	16	20	4.0/6

E150 CLUB WAGON

	L4	14	17	4.9/6
	L4	13	17	5.0/8
	L4	11	14	5.8/8

GMC

G15/25 RALLY 2WD

	L4	15	19	4.3/6	
	L4	14	18	5.0/8	
	L4	14	19	5.7/8	
	L4	16	22	6.2/8	D

G35 RALLY 2WD

	L4	13	17	5.7/8

SAFARI AWD (PASSENGER)

	L4	16	19	4.3/6

SAFARI 2WD (PASSENGER)

	L4	16	21	4.3/6

TOYOTA

PREVIA

	L4	18	22	2.4/4
	M5	19	22	2.4/4

PREVIA ALL-TRAC				
	L4	17	21	2.4/4
	M5	18	21	2.4/4

VOLKSWAGEN

VANAGON SYNCRO 4WD	M5C	16	16	2.1/4
VANAGON 2WD	A3	17	17	2.1/4
	M4	18	19	2.1/4

2WD SPECIAL PURPOSE

CHEVROLET	TRANS	CITY	HWY	ENG/CYL	
APV 2WD	L3	18	23	3.1/6	
R1500 SUBURBAN 2WD	L4	13	17	5.7/8	
	L4	16	22	6.2/8	D
S10 BLAZER 2WD	L4	18	23	4.3/6	
	M5	17	23	4.3/6	

CHRYSLER

TOWN & COUNTRY 2WD	L4	18	23	3.3/6

DODGE

AD150 RAMCHARGER 2WD	A4	13	16	5.2/8
	A4	11	13	5.9/8
CARAVAN 2WD	L3	21	25	2.5/4
	L3	19	23	3.0/6
	L4	19	24	3.0/6
	L4	18	23	3.3/6

FORD

EXPLORER 2WD	L4	16	21	4.0/6
	M5	18	23	4.0/6

GEO

TRACKER CONVERTIBLE 2WD	M5	25	27	1.6/4

GMC

JIMMY SONOMA 2WD	L4	18	23	4.3/6	
	M5	17	23	4.3/6	
R1500 SUBURBAN 2WD	L4	13	17	5.7/8	
	L4	16	22	6.2/8	D

ISUZU

AMIGO 2WD	M5	19	22	2.3/4
	M5	17	21	2.6/4
RODEO 2WD	M5	18	22	2.6/4
	L4	15	18	3.1/6
	M5	16	20	3.1/6

JEEP

CHEROKEE 2WD	M5	19	23	2.5/4
	L4	15	21	4.0/6
	M5	17	22	4.0/6

MAZDA

MPV	L4	18	24	2.6/4
	M5	20	25	2.6/4
	L4	17	22	3.0/6

NISSAN MOTOR COMPANY, LTD.

AXXESS	L4	20	24	2.4/4
	M5	21	26	2.4/4
PATHFINDER 2WD	L4	17	21	3.0/6
	M5	15	19	3.0/6

2WD SPECIAL PURPOSE (cont'd)

	TRANS	CITY	HWY	ENG/CYL
OLDSMOBILE				
SILHOUETTE 2WD	L3	18	23	3.1/6
PLYMOUTH				
VOYAGER 2WD	L3	21	25	2.5/4
	L3	19	23	3.0/6
	L4	19	24	3.0/6
	L4	18	23	3.3/6
PONTIAC				
TRANS SPORT 2WD	L3	18	23	3.1/6

SUZUKI

	TRANS	CITY	HWY	ENG/CYL
SAMURAI 2WD	M5	28	29	1.3/4
SIDEKICK 2WD	M5	25	27	1.6/4

TOYOTA

	TRANS	CITY	HWY	ENG/CYL
4-RUNNER 2WD	A4	19	20	2.4/4
	L4	17	21	3.0/6

	TRANS	CITY	HWY	ENG/CYL
	M5	13	16	5.0/8
	L4	11	16	5.8/8
EXPLORER 4WD	L4	15	20	4.0/6
	M5	16	20	4.0/6

4WD SPECIAL PURPOSE

CHEVROLET

	TRANS	CITY	HWY	ENG/CYL	
S10 BLAZER 4WD	L4	17	22	4.3/6	
	M5	16	20	4.3/6	
V1500 BLAZER 4WD	L4	13	15	5.7/8	
	M4C	12	14	5.7/8	
	L4	16	22	6.2/8	D
V1500 SUBURBAN 4WD	L4	12	16	5.7/8	
	L4	15	21	6.2/8	D

DAIHATSU MOTOR CO., LTD.

	TRANS	CITY	HWY	ENG/CYL
ROCKY 4WD	M5	23	23	1.6/4

DODGE

	TRANS	CITY	HWY	ENG/CYL
AW150 RAMCHARGER 4WD	A4	12	13	5.2/8
	M4C	11	14	5.2/8
	A4	11	13	5.9/8
	M4C	9	12	5.9/8
CARAVAN 4WD	L4	17	21	3.3/6
COLT VISTA 4WD	M5	19	25	2.0/4

FORD

	TRANS	CITY	HWY	ENG/CYL
BRONCO 4WD	L4	14	17	4.9/6
	M5	14	18	4.9/6
	L4	13	17	5.0/8

4WD SPECIAL PURPOSE (cont'd)

	TRANS	CITY	HWY	ENG/CYL	

GEO

	TRANS	CITY	HWY	ENG/CYL
TRACKER CONVERTIBLE 4WD	L3	23	24	1.6/4
	M5	25	27	1.6/4
TRACKER 4WD	L3	23	24	1.6/4
	M5	25	27	1.6/4

GMC

	TRANS	CITY	HWY	ENG/CYL	
JIMMY SONOMA 4WD	L4	17	22	4.3/6	
	M5	16	20	4.3/6	
V1500 JIMMY 4WD	L4	13	16	5.7/8	
	M4C	12	14	5.7/8	
	L4	16	22	6.2/8	D
V1500 SUBURBAN 4WD	L4	12	16	5.7/8	
	L4	15	21	6.2/8	D

ISUZU

	TRANS	CITY	HWY	ENG/CYL
AMIGO 4WD	M5	16	19	2.6/4
RODEO 4WD	L4	15	18	3.1/6
	M5	15	19	3.1/6
TROOPER	L4	15	18	2.6/4
	M5	16	18	2.6/4
	L4	15	18	2.8/6
	M5	15	18	2.8/6

JEEP

	TRANS	CITY	HWY	ENG/CYL
CHEROKEE 4WD	M5	18	22	2.5/4
	L4	15	20	4.0/6
	M5	17	22	4.0/6
GRAND WAGONEER 4WD	A3	11	13	5.9/8
WRANGLER 4WD	M5	18	20	2.5/4
	A3	15	17	4.0/6
	M5	17	22	4.0/6

LAND ROVER	TRANS	CITY	HWY	ENG/CYL	
RANGE ROVER ...	L4	13	16	3.9/8	P

MAZDA

	TRANS	CITY	HWY	ENG/CYL
MPV 4X4............	L4	15	19	3.0/6
	M5	16	20	3.0/6
NAVAJO.............	L4	15	20	4.0/6
	M5	16	20	4.0/6

MITSUBISHI

	TRANS	CITY	HWY	ENG/CYL
MONTERO..........	L4	16	18	3.0/6
	M5	15	18	3.0/6

NISSAN MOTOR COMPANY, LTD.

	TRANS	CITY	HWY	ENG/CYL
AXXESS AWD	L4	18	22	2.4/4
	M5	19	23	2.4/4
PATHFINDER 4WD..................	L4	15	18	3.0/6
	M5	15	18	3.0/6

OLDSMOBILE

	TRANS	CITY	HWY	ENG/CYL
BRAVADA AWD .	L4	17	22	4.3/6

PLYMOUTH

	TRANS	CITY	HWY	ENG/CYL
COLT VISTA 4WD..................	M5	19	25	2.0/4
VOYAGER 4WD ..	L4	17	21	3.3/6

4WD SPECIAL PURPOSE (cont'd)

	TRANS	CITY	HWY	ENG/CYL	
SUBARU					
LOYALE WAGON 4WD..................	A3	22	22	1.8/4	
	M5	24	29	1.8/4	
LOYALE 4WD	A3	22	23	1.8/4	
	M5	24	29	1.8/4	
SUZUKI					
SAMURAI HARDTOP 4WD..	M5	28	29	1.3/4	
SAMURAI SOFT-TOP 4WD	M5	28	29	1.3/4	
SIDEKICK HARDTOP 4WD..	L3	23	24	1.6/4	
	L3	22	23	1.6/4	★
	M5	25	27	1.6/4	
	M5	23	25	1.6/4	★
SIDEKICK SOFT-TOP 4WD	L3	23	24	1.6/4	
	M5	25	27	1.6/4	

TOYOTA

	TRANS	CITY	HWY	ENG/CYL
LAND CRUISER WAGON 4WD	L4	12	14	4.0/6
4-RUNNER 4WD	L4	18	19	2.4/4
	M5	19	21	2.4/4
	L4	14	18	3.0/6
	M5	16	19	3.0/6

CAB CHASSIS

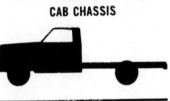

BUICK	TRANS	CITY	HWY	ENG/CYL	
COACHBUILDER WAGON.............	L4	16	25	5.0/8	

CADILLAC

	TRANS	CITY	HWY	ENG/CYL	
COMMERCIAL CHASSIS.............	L4	15	23	4.9/8	GP

CHEVROLET

	TRANS	CITY	HWY	ENG/CYL
POSTAL CAB CHASSIS 2WD...	L3	17	17	2.5/4

DODGE

	TRANS	CITY	HWY	ENG/CYL
DAKOTA CAB CHASSIS 2WD...	L4	12	12	3.9/6
	M5	12	16	3.9/6
	L4	13	17	5.2/8
D250 CAB CHASSIS 2WD...	A4	11	13	5.9/8

TOYOTA

	TRANS	CITY	HWY	ENG/CYL
CAB/CHASSIS 2WD..................	L4	11	11	3.0/6
	M5	11	12	3.0/6

Index